Tomorrow's World Looks to the Eighties

Edited by Michael Blakstad

British Broadcasting Corporation

Published by the British Broadcasting Corporation,
35 Marylebone High Street, London W1M 4AA

ISBN 0 563 17597 4

First published 1979

Printed in England by Jolly & Barber Ltd, Rugby

Contents

This book, like the Tomorrow's World programmes on which it is based, would not have been possible without the continuous and imaginative work of the producers, researchers, directors and many others who operate behind the scenes. We, the writers, are glad of this opportunity to thank them.

Michael Blakstad

Lifestyle

Build Yourself a Future

Because its topic is the home in the 1980s, this is a do-it-yourself chapter. To create your own vision of the everyday life in the next decade, take all or some of the following possibilities and mix according to taste.

A world-wide network of telephones and television screens linked both to homes and to computers, making the communication of information and of visual images a flexible activity in which anyone can join.

A rash of microprocessors, the mighty minis which will appear in corners of the home where the electronic giants of the 1960s would have proved too cumbersome and too expensive to contemplate.

Less work to go round, and therefore more time on people's hands, created both by the automation of many repetitive jobs and by the education of many people in developing countries to whom industrial jobs had previously been denied.

A mounting concern for some of the earth's dwindling resources, and particularly for liquid fossil fuel. Britain, in the 1980s, will still be in the middle of her own North Sea gusher, but several traditionally oil-rich countries will be running out of theirs, and it is possible that the deeper offshore fields and the shale deposits may prove too expensive to exploit. Whether the imminent shortage is real or imagined, social conscience and government action may well be forcing families to conserve oil and other non-renewable resources.

Which possibilities you choose will depend on your estimation of their likely importance. How hard, on the one hand, are scientists and engineers going to work at developing the relevant technologies and skills? How much is the consumer, on the other hand, going to encourage or even demand action? The glimpses which are offered here of 'tomorrow's world' are descriptions of the technically possible – in many cases achievements which have already taken place – and it is up to the reader to decide how likely it is that use will be made of them in the decade ahead. Let us start very much on home ground, with the future of television.

Two-way Television

The technology exists – this phrase will occur again and again throughout the chapter – to transform television from its present one-way, transmitter-to-receiver form into an active, two-way mode of communication.

On 1 December 1977, a new television station opened in Columbus, Ohio, which contains many of the elements from which future television services will be constructed. Columbus is a pleasant mid-Western town, home to Ohio State University which has not only one of the best college football teams in the United States but also a variety of cultural and scientific facilities which blend happily into the town's own lifestyle. There's the Battelle Institute, one of the most productive research institutes in the world, the place where Xerox technology was developed and successfully launched. And now, Columbus also houses QUBE, Warner Brothers' latest contribution to the cultural life of the nation.

QUBE is housed in a converted supermarket just on the fringes of the university campus. Its whitewashed walls contain one of the best equipped small television services in America; three fairly sizeable studios, a number of one camera 'mobile' units, and most important of all, a computer. The computer sits in a glass goldfish-bowl in the heart of the building and it is linked by telephone line to each of the 100,000 subscribers in Columbus who have paid $7.50 a month to plug in to their very own television service. The computer monitors exactly which of the thirty available channels each family is watching while the family has a small control panel with separate buttons for each of the stations. By pressing any of the thirty channel selector buttons, the viewers can switch programmes – nothing new here, except the number of channels available – and by pressing one of five specially numbered extra buttons at the side of the panel, they can communicate their views to the computer and to the people making the programmes.

Those thirty programmes come in three categories. On the left of the console are the ten 'premium' channels: pay TV-services like first-run movies, big sporting events, concerts and so on. When the viewer tunes in to any of these (he's allowed two minutes' sampling time to decide whether he wants to continue) the computer silently tots up the bill and prepares to send to that home an account to cover the cost of 'premium' viewing. The right-hand side of the panel presents the Columbus viewer with access not only to the four channels available to anyone in that city, but – via cable – to the output of other stations in cities normally outside the range of their television antennae, Cleveland, Indianapolis and Cincinnati, for example. There is no extra charge for this service.

The middle line of buttons plugs the family in to 'community' programmes, and it is here that QUBE believes it is pointing the way for the future of television. Most of the buttons simply provide a very detailed information service – mainly in teletext form – tailored to special interests like sports news, consumer information (the station checks prices at all the main supermarkets on a day-to-day basis), education, stocks and business news and so on. But the top button is marked Columbus Alive; press this, claims the publicity, 'and enter the era of 2-way participation in the infinite, unfolding, never-ending worlds of QUBE'.

This is where the five numbered buttons come into their own. One of the first

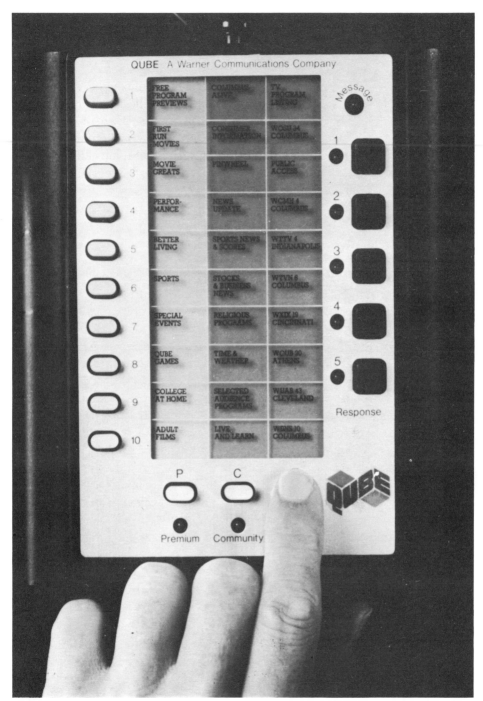

At the press of a button . . . enter the Wonderful World of Qube

programmes transmitted on *Columbus Alive* was called *How Do You Like Your Eggs?*, in which a hearty, check-jacketed link-man asked his television audience to 'Press button One if you think Nelson Rockefeller is Vice-President of these United States, button Two if it's Billy Carter, Three for Walter Mondale . . .' and so on. For the record, 64 per cent of QUBE subscribers knew the answer to that one. Another first-night show was *Going Once, Going Twice* and featured Flippo the clown encouraging viewers to press their response buttons if they wanted to buy what he had on offer. A third was *Columbus Bananaz* in which viewers frantically pressed high or low buttons according to their like or dislike of the performer on screen at that moment. When the rating fell below 50 per cent, off went the star! Even straightforward items of community news were punctuated by invitations to the viewers to punch their buttons if they wanted to know more about the subject under discussion.

If the early offerings of the world's first two-way television service were anything to go by, the future of this kind of 'access' would not look rosy. In the first week the entire station was hypnotised by the power of those tiny buttons. The programmes were dominated by the need to include an issue which could be thrown open to the five buttons, to a multiple-choice question which would prove to the viewers that their role in the content was vital, and to the station executives that two-way participation was alive and well. The programme standards were roughly those of a developing country which had just opened its first television service: despite the fact that the technical equipment was brand new and very expensive, and the staff recruited from Los Angeles and New York, there appeared to be a grim determination that gloss and old-fashioned professionalism had no part to play in access television. At the same time, in complete contradiction to this philosophy, the blazer-clad presenters of the interview shows and the performers – like Flippo – imported to enliven the family programmes had brought with them the values of highly commercialised television. They tried for all they were worth to bring the humdrum nature of their subject matter and interviewees up to a level of entertainment which would have satisfied the moguls of Hollywood and Sixth Avenue.

There is, of course, what the Americans call an 'up-side'. Some youth discussions, allowed to run infinitely longer than would be possible on any other American channel, did provide a sense of involvement; they were so absorbing that the presenters forgot to refer to the buttons. Guitar lessons and experiments in extrasensory perception would never normally find their visual way on to Columbus screens, never mind *Madame Butterfly* and Bach's *Magnificat*. But the reason for analysing such a newly-founded experiment is to ask those who believe that community programming will be the salvation of our television-orientated civilisation precisely what kind of service they have in mind.

Telecommunications (see pp. 115 ff.) have made it possible to link houses to television stations, to computers and to each other in an almost infinite series of different combinations. If, as is very likely, the future is going to diminish our opportunities for social contact, then these communication links are going to become very important indeed. There should, in theory, be plenty of opportunity to increase the level of public debate and understanding on many different aspects of tomorrow's world, both by giving people access to more banks of information, to experts and to events, and then by encouraging them to use the television network as their forum for debate.

But what has happened so far in Columbus is that the station has found itself caught by the pressure, on the one hand, to establish itself as a viable television service, resorting to many of the least attractive elements of American programming, and on the other hand has been seduced by its own technological prowess in allowing the computer and its results service to dominate the programme service. In Columbus, five buttons rule, and it's not OK.

The Mighty Minis

'You will awaken some morning five years hence, speak a few simple words from your bed to your toaster, coffee pot and frying pan, and walk into the kitchen fifteen minutes later to a fully prepared breakfast.

'The same computer that is wired into the walls of your house and built to recognise your voice will turn on lights when you walk into the kitchen and turn them off when you leave. It will also turn the refrigerator off when you leave for work and turn it on before anything defrosts.

'Your furnace and air-conditioner will respond to the same computer, driving warm or cool air into only the rooms that are occupied. Your home computer will pay your bills and figure out your checking account balance. It will order your groceries, plan your meals and suggest recipes for dinner guests on fussy diets. It will even open the front door for you, responding without locks and keys to the sound of your voice.'*

And so it goes on. These were the predictions of an American journalist writing in 1977, and there have been dozens like them, stretching back for at least a decade. Until very recently, such prophesies have been treated with healthy scepticism; the computer revolution has been many times hailed but remarkably sluggish in showing itself. Now, however, one fact is turning the tide in favour of such fantasies arriving, fully functional, on our front doorsteps – perhaps even sooner than the five years allowed by Thomas O'Toole – and that is the now-familiar phrase: the technology exists. . . .

Back in 1972, a USA trade weekly, *Electronics News*, announced without

* Thomas O'Toole, *The Washington Post*.

enthusiasm that Intel Corporation and Texas Instruments had won a tender from Computer Terminal Corporation of San Antonio and were about to put into production something known as a processor on a chip. In fact, Texas Instruments itself seems to have been unaware of the significance of the event, because it withdrew from the project leaving Intel to produce the world's first general purpose microprocessor. It made Intel into a technological giant, very quickly, even judged by the hot-house standards of California's Silicon Valley, and today, every country in the developed world is trying to establish its own microprocessor industry. In 1978 Britain attempted to make up for its failure to generate a home-grown microprocessor industry, with the NEB spending £50 millions on the launch of INMOS (the brainchild of an American entrepreneur) and with GEC setting up in partnership with the American company, Fairchild, and with the critics of both claiming that this represented too little too late. The Japanese, needless to say, are saturating Palo Alto with their spies, making use of the Americans' generous offers of 'factory visits' to observe and where possible imitate the secret of American success.

It is, of course, the integrated circuit which lies at the heart of the automated lifestyle of the future. It has reduced the cost of computing power to a level which anyone can afford, and the size of today's microprocessors and mini-computers to a scale which tucks them snugly into the most compactly designed implements or living spaces.

The challenge faced by Intel and now by dozens of other manufacturers of integrated circuits is that of compressing as complex a circuit of electronic pathways into as small a space as possible. The material which makes this possible is silicon (sand has the advantage of being a resource which will never run short). Manufacture starts with crystalline silicon of 99·9 per cent purity, lovingly 'grown' by specialists somewhat in the manner of rock candy; the property of this material is that it can be made conducting or non-conducting according to the way the manufacturer treats it. The way to make some parts of the chip electrically positive and therefore capable of conducting signals is to 'dope' pathways in the silicon with impurities; negative zones are created by charging the rest of the silicon with electrons. What happens is that the logic designer draws up a large-scale photomask of the network of commands which the chip is to perform; the mask is then reduced photographically to a microscopic scale. This mini-mask is then placed over the carefully prepared silicon and the network is etched into it by exposure to ultra-violet light; the areas which are shielded from the rays remain soft and are then washed away in acid while the areas which have been exposed to ultra-violet harden and resist the acid.

Electron and laser beams are now being developed to improve the making of these vital masks, while some factories use ion beam implantation to 'shoot' atoms of impurity directly into the silicon material: that's how fine the techniques are

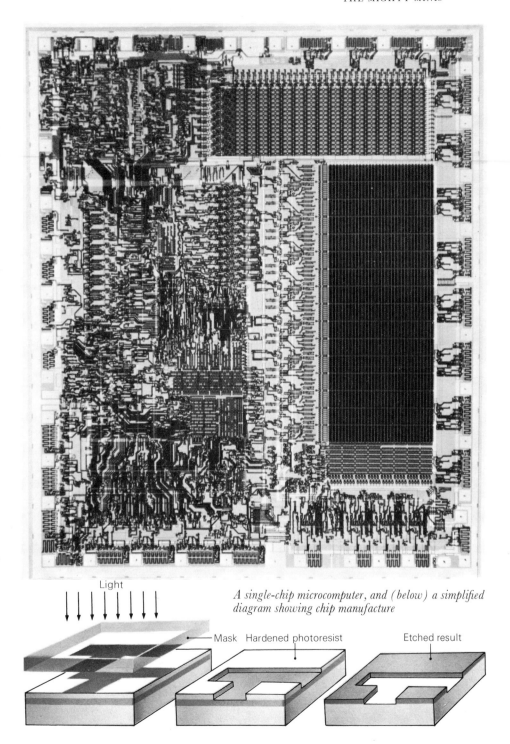

Light

Mask Hardened photoresist Etched result

A single-chip microcomputer, and (below) a simplified diagram showing chip manufacture

becoming. A 3 millimetre by 3 millimetre wafer of silicon can house well over a million 'gates', capable of performing the number of calculations which a mere decade ago could only have been carried out by a hall full of computer power – and now they are selling at less than £5!

But producing the integrated circuit is not the end of the story; you have the computing power, but how are you going to use it? Well, this is where most microprocessors set off for your home along one route, while the home computer sets off along another. In a way, the difference between the two journeys can be illustrated by an analogy.

Say a composer has just created a masterpiece of music which he is anxious to share with a friend in another town. He could easily ring his friend, place his own telephone next to the piano in his house, and play the music. This is what the self-contained microprocessor usually does: it receives the signals, in this case the music, directly down the line in precisely the form it left the composer. But say the friend is also a fine musician, and capable of responding very rapidly to the composer reading the notes to him and simultaneously playing the music on his *own* piano. In this case, the composer has broken the music waves down into code form, and the friend is acting on what, in computer terms, is a digital form of instruction. This is how the computer behaves: the analogue signal (the music) is broken down into digital form (the notes) and acts on these coded instructions. (Michael Rodd's chapter on Leisure gives a more direct explanation of analogue and digital.)

Now, the microprocessor which has said goodbye to the factory and set off on the more direct route to your home is designed to receive its instructions in analogue form (like the music). It can respond to the output of a microphone, for example, or that of a heat detector. It is simple to plug this kind of microprocessor directly into, let's say, an oven, and it will be able to respond to changes of temperature inside the meat and control the cooker so that the roast is crisp on the outside, still pink in the centre, just the way the cook wants it to be. That's because the logic designer has given this kind of microprocessor a tiny analogue-to-digital function built into its circuitry; all the makers of the cooker had to do was to connect one input of the microprocessor to a temperature-sensing device, connect another terminal to the console on which the cook selects from the tasks the circuitry can perform, and then connect the output of the microprocessor to the controls of the cooker. If the integrated circuit has been accurately designed and made without the tiniest speck of dust or unwanted impurities in the network, the microprocessor will do the rest.

The jobs being done by microprocessors go way beyond roasting joints to the correct turn. Airliners today carry them in their undercarriage to detect any change of friction which might lead to a skid; the slightest hint of trouble and they automatically release the wheel locks to prevent a disaster.

A tractor exists which is actually driven by one. It can sense the turns and furrows of the field and follow them up perfectly. In America, there is a motor manufacturer who has a microprocessor in his car. It is activated when the owner leaves the car and it sits there on the alert for careless drivers of other vehicles. If it detects the danger of another car backing into it, it releases the brake, starts the engine, and backs out of danger.

In the home, the applications may be less dramatic, but they're more universal. Attached to a simple smoke detector they can warn the fire brigade and set off the correct extinguishers. Temperature control, with subsequent savings in fuel bills, is simple to achieve, and microprocessors are built into many motor cars, to monitor fuel consumption or to control exhaust emissions. TV games are the creation of the microprocessor, so are the latest TV tuning devices, and there's at least one inexpensive camera whose microprocessor takes care of all the shutter speed and exposure settings. At least half the jobs in Thomas O'Toole's electronically controlled house of the future are routine functions which today's microprocessors would have no difficulty performing.

Other tasks, however, can only be performed if the range both of instructions and of calculations is much wider and more complicated than this kind of microprocessor, with its limited forms of input, is capable of completing. This is where another form of microprocessor comes in, one capable of performing more calculations of greater complexity, but which can only cope with digital instructions. It can't function unless it is linked to a more powerful memory, to a power supply, and to at least one 'interface device' – in other words, some means by which it can receive instructions and transmit its replies. This microprocessor is destined not for a car exhaust or a cooker, but for a computer.

Until recently, computers have been manufactured only by specialist companies, for two reasons. In the first place, the really difficult part in making computers lies in designing the system; the sort of computer which is going to handle the reservations of a complete airline, or the invoices of a major firm, requires systems design of the order which can only be achieved by professionals. Secondly, the business of assembly has always involved both special equipment and skilled wiring, which could only be achieved by an experienced workforce. The arrival of the microprocessor, with the logic and the wiring built in, so to speak, at its birth has changed all that.

Today, if the tasks required of the computer are relatively simple, there is no reason why anyone with an elementary knowledge of logic and a deft hand with a soldering iron could not manufacture his or her own computer to perform exactly those jobs around the house which match his or her lifestyle. The advantages of doing-it-yourself are twofold: the traditional one that the components come cheaper and you are not paying for labour and manufacturer's overheads; the additional one that your computer is designed for you personally.

In 1977, a British businessman called Gordon Ashbee gambled £40,000 of his own money on the hunch that this country would witness a boom in do-it-yourself computers. He had observed the rapid proliferation in America of shops selling computer components – some streets in Los Angeles boast not one but two or even three such stores – and he also believed that the American manufacturers were unable to supply the assembly kits to their home markets, never mind export them.

So Ashbee set up his own small factory in Hertfordshire where, borrowing circuit designs under licence from an American company, he collected the different components – keyboards, central processors, disc or cassette memories, video displays and the rest – and employed a handful of girls to assemble them in kit form. He also set up, in London, the first of what he hoped would be a chain of shops through which he could sell the kits, whilst attempting to persuade other retailers – like hi-fi shops – to sell his products.

In many respects, the do-it-yourself computer business looks like following the course set by hi-fi. That craze started at the do-it-yourself end of the market; the enthusiast who knew enough about woofers and tweeters and diaphragms to go into a component shop and come out with an armful of selected loudspeakers, gram decks and pick-up arms started a fashion which was only slowly followed by the gramophone manufacturers. Even when the first complete hi-fi sets arrived on the market, they were so much more expensive than the sum of the components (and, in any case, infra-dig for the genuine enthusiast) that the components business is still alive and well.

The home computer business is following the same pattern – but more rapidly and with a smaller price margin between the kit and the complete package. In early 1978, the Americans arrived with ready-made packages costing as little as £200, with British manufacturers hot on their heels. The tasks which could be performed by these low-cost units was, admittedly, extremely limited: even £200 is a lot to pay for a mini-computer which can only give you the option of playing a variety of electronic games and storing your household accounts, but the threat to Ashbee's business was apparent.

In fact, as happens in this kind of market, the arrival of the enemy appeared to have had the opposite effect. Far from mopping up the bulk of a small market, the fresh impetus seems to have increased interest and with it the number of potential buyers. Ashbee moved his factory from the 800-square-metre building in Hertfordshire to a 3000-square-metre plant near Peterborough. His turnover for the first quarter of 1978 was £200,000, and he opened a second shop – under franchise – in Manchester. His claim – as with the hi-fi buffs – was that the over-the-counter computer is too restricting, the customer can only buy the services offered by the manufacturer and cannot design the computer to suit himself, whereas the do-it-yourselfer can build the system as he or she pleases. Above all,

Ashbee hoped he had taken the myth out of computer manufacturing – the myth that only the advanced electronics engineer could dream of assembling one – and in doing that he helped create and enlarge his own market.

The home computer boom really does seem to have caught on in Britain, but sadly, Gordon Ashbee will be playing no part in it. In the early months of 1978, he received what he described as 'an offer he couldn't refuse' from the very people he couldn't afford to oppose – the American company from whom he licensed the circuit designs. In April 1978, he was forced to quit his original computer workshop, and the new bosses promptly closed it down to open new branches in other parts of London. One condition of the deal was that Ashbee should in no way become involved in the home computer business for a year. After that, well, a year is a long time in microprocessors.

Time on Your Hands

A wry comment on the last topic is the story of the microprocessor firm which was doing so well that it had to move into smaller premises; the clear message from the future is that there will be less work to do – and therefore less employment. Sid Weighell, General Secretary of the National Union of Railwaymen, was being both serious and realistic when he predicted on *Tomorrow's World* that his members would soon be working a twenty-four-hour week; computerised signal boxes and driverless freight trains, both technical possibilities already, would inevitably have two effects on railwaymen: they would need a higher degree of training – their jobs would evolve into something closer to air traffic controllers than today's manual signalmen and drivers – and there would be less work to go round. The implication is that there would either be massive layoffs, or Sid Weighell's shorter working week.

For the railwaymen you can substitute almost any industry in Britain. Not only will automation reduce our manning levels, but the growing industries in Third World countries, with their lower wages, could cut harshly into Britain's share of the world market. One million, four hundred thousand unemployed people are already causing enough problems, and if jobs continue to shift away from the less educated members of our population and towards the technological elite, less work could mean a huge class of unskilled unemployed, with an inevitable spate of vandalism and delinquency.

There is an altogether more optimistic future in store if education (see pp. 123 ff.) succeeds in tailoring students more precisely to the kind of jobs that will be available when they leave school and in doing so, spreads the available employment more evenly across the population. In Sid Weighell's scenario, this means less unemployment, in terms of the total number of people actually without jobs, but fewer hours worked by each individual and consequently more time spent at home.

For this reason, one of *Tomorrow's World*'s earliest foibles is already stillborn. Back in 1966, the programme featured a charming robot designed by the ever-ingenious Professor Meredith Thring and affectionately named Mabel. 'Begin the day with able Mabel, she'll run your bath to exactly the temperature she knows you like, she'll make your bed and lay out clothes for yet another day, free from household drudgery, free from time-wasting chores like vacuum cleaning.' Somehow, Mabel has not risen from the drawing board; it is not just that she defeated the efforts of Professor Thring to design the controls necessary for such an ambitious lady; the fact is that labour-saving devices have slipped way down society's list of priorities. In a world when time and effort are very possibly going to be over-abundant, domestic robots are almost certainly a thing of the past.

Of greater concern, will be the need to conserve resources. Although a later chapter suggests many alternative methods by which energy could be generated, and although the eighties will witness the peak of Britain's offshore oil production, world supplies of oil will probably be dropping, while the price of all forms of energy will certainly be rising, so it is not impossible that the name of the domestic game in this decade will be to use human resources in order to conserve, not only energy, but a lot of other natural resources.

The amount of *waste* which is discarded without any attempt at recycling, pricks the conscience of the conservationist, and here is one area where more work by the householder can, in theory, save resources. Take glass for example. The use of broken green glass can actually improve the manufacture of new glass, but since the raw materials of glass – limestone, sodium carbonate and sand – are as abundant as the grains in the Sahara desert and since there is thought to be a limit to the amount of new glass which can be made with cullet, no one has really paid much attention to the possibility of recycling this broken glass.

Not, that is, until a research group at Cardiff University put its thinking cap on and worked out that it takes less energy to melt old glass than to make new: boost the percentage of used cullet, it says, and it will save money and resources. It went further. Powdered glass, suitably treated, can improve the properties of some plastics and reduce the cost of making them. Waste glass mixed with sump oil (another waste product) and asphalt can be used for road repair materials. The group has experimented with bricks, building panels and tiles containing a high proportion of powdered glass. It is even looking at a possible use as a mulch for controlling weeds, and as a replacement for crushed stone in chicken grit.

The Cardiff researchers' proposals depend, of course, not only on their experiments proving to be technically sound, but on the economics of separating out the waste and collecting it. Lorries use up fuel just as furnaces do. That has always been one of the keystones of the argument used by drinks manufacturers in favour of introducing one-trip bottles. Whatever the environmentalists say, claim the makers, our filling machines can only fill our own bottles. Other brands have other shapes; it is all part of marketing, and lorries would have to make long, expensive journeys collecting only a few bottles from each collection point. Supermarkets in any case will not give up the space needed if their customers are to return their bottles, but the strongest argument of all is levelled against the consumer: sale-or-return bottles have to be made of heavy-duty glass, and the cost of this glass requires at least eight return journeys to justify the expense. In fact, although the Scots seem capable of maintaining an average of around sixteen trips per bottle, the average British figure is way below the necessary eight: in London the drinks people can only expect to see a bottle back three times; in seaside towns and other day-trip havens, the figure is even lower than that.

It is the consumer, once more, who is blamed for the amount of wood pulp Britain imports to make paper (and for the trees cut down to provide the pulp). Our import bill runs at well over £1000 million a year, but the paper industry refuses to boost the amount of waste paper it uses in the manufacturing process because, it says, the public would not put up with greyer, coarser newspapers produced with recycled fibres. An American experiment is producing a paper made with 100 per cent second-hand fibre; Britain's newsprint does not go above 20 per cent. If all Britain's newsprint were made from recycled fibres that would use over 380,000 tonnes of waste paper a year, and would save a lot of trees.

It would not necessarily help the British balance of payments, however, if the present distribution system were not improved to catch up with the trend. At present, Britain is setting new records for the waste paper it imports for use in the manufacture of cardboard. In 1977, 6 million tonnes of British waste paper was incinerated – the results of the enthusiastic but ill-organised spate of paper collection which had left many a scout hall full of the stuff – while shiploads of foreign waste paper were arriving at our ports.

It is a perfect vicious circle; no one will set up the collection system until the market is firmly established, while there is no real market for paper made with re-cycled fibre both because the cost is too high and because the consumer has little chance to demonstrate his or her choice.

The town of Kirklees, near Huddersfield, has shown the glimmerings of a distribution system which could justify the recycling both of paper and of glass. Six thousand households there have experimented in an Oxfam 'Wastesaver' scheme in which each home was given a segmented rubbish holder, clearly-labelled lids opening to five different plastic bags reserved for different categories of waste: paper, jumble, tins, glass, and plastic. Paid collectors came for the rubbish, keeping it separate, and delivered it to a depot run by volunteers where the saleable waste was despatched; long-term contracts had been struck for the paper, the tinplate and the glass, while the textile industry was delighted to receive rags to buffer it against the decline of the traditional totter. Enthusiasm has been high at Kirklees, and another scheme in Switzerland (now being tried in several towns in Britain) introduced the idea of large skips with three separate compartments. These are designed for the collection of old bottles (green, brown and clear glass), and the so-called 'bottle-banks' are left in supermarket car-parks, ready for the shopper to dump the empties and for the council to collect and sell the glass. Results, in 1977, were ecologically encouraging – 1 million bottles collected in Oxford alone – and, what is more to the point, profitable. If, in the eighties, the householder is going to have a lot of time to fill, perhaps some of it will go towards separating out the rubbish. Once again, the technology exists for re-cycling not only the paper and the glass, but also the tins (automatic scouring – Warren Springs laboratories), the plastics (Widnes has a plant which can pro-duce a new thermoplastic material re-using old plastics) and, perhaps the most shameful waste of them all, the vegetable refuse. That technology is called 'composting' and a 1976 Government report drew attention to the fact that very few kitchens separate out their organic waste for use in the garden. Strange, when gardening is reckoned to be Britain's national pastime.

Cooking is another activity which, at present, wastes a lot of energy. In a standard cooker, 60 per cent of the power is wasted, much of it because it escapes through the sides and the top, both of the oven and of the saucepans on the hotplates, and a lot more because it is generated at wavelengths which have no effect whatsoever on the food. The cosy red glow of an electric ring or of a gas oven may have a psychological effect on the cook, but it has no physical effect on the food. The work of cooking is done by invisible rays occupying about a third of the infra-red part of the spectrum; the other wavelengths represent wasted energy.

Microwaves, of course, seem to be a very efficient way of cooking. They operate by focusing on individual moisture molecules in the food and agitating these molecules very rapidly – 2000 million times a second – so that the moisture molecules then agitate those around them and the heat generated by all that activity then spreads very rapidly indeed. That is why a glass can remain cool in a microwave oven, while the water inside it boils in less than a minute, and why the air of the oven doesn't warm up at all. So could microwaves be the end of the energy problem? No, not quite. There are two major drawbacks to microwave cooking. The beam is inherently extremely dangerous. Subject your hand, or worse still, your brain, to that agitation and it will be comprehensively addled. A high proportion of the cost of a microwave cooker goes into safety precautions, yet doubts persist; in America there have been continued congressional hearings as to whether there isn't still some danger inherent in using microwaves, so there will

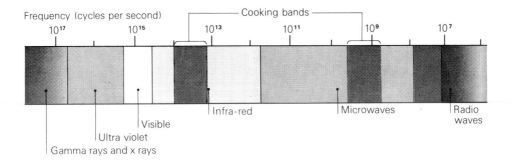

Frequency (cycles per second)

Cooking bands

10^{17} 10^{15} 10^{13} 10^{11} 10^{9} 10^{7}

Infra-red   Microwaves   Radio waves

Visible

Ultra violet

Gamma rays and x rays

Only two parts of the spectrum are useful for cooking food: a portion of the infra-red sector, and micro-waves

always be some who would prefer us to avoid this particular technology. At a less serious, but in the long run a more telling level, there is the problem that microwave cooking doesn't brown or crisp the outside of the food. This has two causes: the air in the microwave oven is not hot – as we have observed – whereas the conventional cooker produces the succulent exterior because the hot air cooks the outside of the food first and the heat is then conducted into the centre; secondly the moisture molecules on the exterior, though they are being agitated as vigorously as the others, have only half the number of neighbours on which to work, and consequently they only get half as hot. The result is chicken which looks, and tastes, slightly parched on the outside, sausages which are both limp and colourless, and toast which is nothing more than hot bread. Some microwave ovens have browning units added, but few homes rely totally on this form of cooking.

A group of engineers in the Midlands, an enterprising collection of researchers attached to an engineering and foundry group, is trying to make cooking more effective by sticking to the conventional bandwidths, but concentrating on the range of infra-red which actually does the work. They are experimenting with infra-red elements tuned to the correct frequencies, using glass saucepans and pots in order to let the rays get at the food. The infra-red is certainly efficient – a litre of soup can be brought to the boil in less than two minutes – and it presents an interesting challenge to the cook. Instead of controlling the rate of cooking by simply turning down the power (though this is certainly one way of doing it) the researchers are working on the possibility of sending out continuous rays when the food needs quick browning, and pulsing the rays at intervals in order to cook the food more thoroughly on the inside. Thus, a fish finger is best cooked by an initial burst of continuous infra-red lasting about thirty seconds, then a slow pulse for a minute to penetrate the centre, then a final continuous thirty-second burst to brown the exterior. The whole fish finger programme, as well as the 'don't let the milk boil over' control and all the others, can be stored in a microprocessor to help the harassed cook, and there's also the possibility that infra-red elements could be built into microwave cookers to give the best of both technologies.

But perhaps the best design factor in the infra-red cookers is one which could easily be built into the most conventional oven. The designers found it convenient, in their very first prototype, to arrange the infra-red elements in a vertical circle, forming a heat-well into which the glass saucepan was then lowered to receive the heat from all sides. When they turned off the power, they were surprised to see how very slowly the saucepan lost its heat, insulated as it was from every aspect except the top. Perhaps, long before infra-red cooking becomes available on the market, the manufacturers of conventional cookers will see fit to incorporate this energy-saving heat-well into their hot-plates.

Food could well be a creative outlet for the houseperson in the eighties seeking to occupy increased leisure time. There's unlikely to be any shortage; by the time we've munched our way through the American grain silos, the European Beef and Butter mountains (to say nothing of drinking our way through the Wine and Milk lakes); by then, multi-billion-pound schemes to tap the agricultural resources of the Nile Valley and other under-used fertile areas will surely have come to fruition. The food will, as it should, be more evenly divided amongst the world's population, but at the worst we'll be in the hands of the chemists who will have started to foist on humans – as well as on animals – their single cell proteins. There is plenty of nutrition in algae, but not a lot of taste or colour. The challenge to the cook of the 1980s will be to put fun into eating. No more trips to the supermarket buying double-cellophane-wrapped convenience foods; now the job will be to husband herbs and spices and to develop cooking techniques which will not only save energy but will make vegetable proteins and chemical substitutes taste and look palatable. We've already witnessed microwave cookbooks and health food cookbooks – how long before the first Single Cell cookbook?

Water is another resource which we use all too lavishly. Hosing plants with gallons of water which have been expensively treated to make them utterly safe for humans to drink is utterly wasteful, especially when the plants probably don't want the water in such sudden, intermittent doses. The alternative, which costs more in the early stages and certainly occupies some more of that available time, is 'nutrient film technique'. Flowers and vegetables can be grown in plastic tubes, their roots hanging in a mixture of water and chemical nutrient, so that both water and food can be used and used again until they are absorbed by the plant. Neither food nor water is wasted, and the plants grow more quickly and in less space than conventional beds. The end result is not a sight for evening strollers, but it makes much better use of the resources.

Houses could well have their own small water-cleaning units which, while not approaching the standards supplied by the water boards and definitely not producing water fit for drinking, could nonetheless take some of our bath or dish-washing water and restore it to a high enough standard to be used again for washing humans or dishes or cars or floors. 'Flocculation', for instance, is a technique of pumping air bubbles into the bottom of a tank of dirty water. The bubbles, as they float towards the surface of the tank, attract any particles in the water and carry them to the surface, where they form a scum which can be skimmed off as the water leaves the tank.

There is, of course, no incentive for any householder to install such a plant when there is no penalty for using too much water. In the Californian drought of 1977, householders were fined if they went above their ration; even in normal times they are charged for the amount of water they use because that water is measured by meters. Here we are back in the do-it-yourself prediction game. The business of installing meters in Britain would cause the disruption and expense roughly equivalent to the feeding of natural gas into every home in the early 1970s. Will the need ever justify the means? Britain has plenty of water falling from above, but the drought of 1976 showed just how little space we can afford to store the water we need and how thin the line is between drought and plenty. All that this chapter can suggest is that the technology exists to make much better use of our resources if the powers-that-be decide to make use of them.

The Office of the Future

When the last girl gets the sack and the old typing-pool is finally closed down, the chances are there will not be anyone at the office to take note of the event. The trend which is leading towards this sad event is, however, already starting and it could, in the long run, have as great an impact on our everyday lives as any so far recounted in this chapter.

The reasons behind the word processing revolution are straightforward. The process of dictating, drafting, and finally typing and mailing letters is expensive: estimates vary, but £3.00 is not far from the average mark. The job of taking dictation and typing letters is usually reckoned to be not only tedious but, to the fair sex, positively demeaning; why are there so few male secretaries? Finally, of all today's professions, secretaries are among the least organised and militant when it comes to trade unions; thus divided they could easily fall.

Then there's the slightly wider economic picture. The whole work pattern which requires millions of office workers to troop at the same time to work and back again at the same time of day is extremely expensive in terms of transport costs. Railways and bus companies have to operate enough rolling stock to carry the huge flood of commuters when they only occupy the trains and buses for a couple of hours morning and night; for the rest of the day, the transport stock and staff are an expensive luxury. The roads are jammed with private cars. If only we could abolish commuting. . . .

Well, we could; the technology exists, thanks to a very powerful industrial drive towards computerising office services. Word processing replaces paper with electronics, the typing-pool with a small number of keyboard operators, the secretary of old with a dictation machine at the end of a telephone line, letters with video-displays. Word processing is backed by some of the mightiest names in industry: financial sponsors like Exxon and Citibank, technical innovators like IBM, Xerox, 3M, Philips and a host of others, including a number of small companies with a staggeringly fast growth performance. Alas, hardly one of the innovating firms is British.

In the 'office of the future' (which already exists in a number of large corporations on the other side of the Atlantic), there is no hubbub of typewriters and of secretaries answering the phones, swapping gossip or taking dictation; instead, an unmanned electronic switchboard receives dictation either directly from the person who wants to send the letter or from his or her dictation machine linked to the phone. Still without human intervention, the switchboard interrogates all the consoles in the word processing division and ascertains which operator is best suited to cope with the job.

For the first, and last, time a human appears in the near-deserted office suite. It still needs an operator – his or her skills are a little above that of a typist – to key the message into the computer, though the day will surely come when the executive has his or her own keyboard at home and is required to key in his or her own letter or memorandum. As the script is entered into the computer's memory – floppy discs or cassettes are the most popular forms in use today – the words are displayed on a video display screen; the keyboard operator can alter mistakes simply by pressing a button; no messing with erasers, no need to re-type a whole document because the boss isn't satisfied. If the executive wants the layout changed, or the insertion of a new paragraph, then the whole text can be electronically juggled to meet his or her whim without any of the correct words being erased. The all-forgiving computer simply makes room in its memory and deletes, adds, or justifies whatever is needed. When everyone is happy, then – for the first time – paper appears, spewing out of a high-speed printer at the rate of sixty characters a second, as many copies as are needed, with minor alterations on every one if that is what the boss wants.

Not surprisingly, Britain's first systems are being bought by legal and insurance companies which issue large quantities of very similar documents, contracts or policies.

The next step in the word processing saga comes when the letters are loaded in to an electronic hopper capable of reducing the image on the printed page into

Word processors don't use paper at all until the electronically stored text has been checked and corrected on a visual display unit. Characters are finally printed at high speed by an ink jet

digital code, then transmitting that code down the telephone line to any city or country with a suitable receiver on the other end of its phone. The French Government is so interested in this form of facsimile transmission that it is considering installing it to replace the bulk of its present postal service, enabling it to bypass the whole awkward business of carting paper up and down the country in trains, planes and vans. But it could be that facsimile transmission itself will be outmoded when the telecommunications revolution gets properly under way; after all, when everyone has a computer transmitter-receiver in the home, why commit the message to paper at all? Once the keyboard operator has entered it into the electronic memory, then it will only be a matter of summoning up the text on to your video screen whenever, and wherever, you want it. But – massive though the impact of this will inevitably be on the home lives of at least half the population of Britain – it's a subject which really belongs to a later chapter.

Judith Hann

Less of an Art, more of a Science

Medical progress is expected to be dramatic in the remaining years of the twentieth century. The mood is one of optimism as scientists begin to get to understand many of our more serious diseases. Treatment is gradually becoming less of a hit and miss affair, and more of a scientific problem which can be solved.

During the next twenty years the most optimistic medical prophets believe that we will have coped with the two big killers – cancer and heart disease – and that doctors will have learned to regenerate organs and alleviate ageing. But, while research into these important, dramatic areas progresses, it is essential that the more mundane medical subjects are not ignored.

This is a very real danger, but the Government is aware of it. A study which has already been commissioned will compare investment in research on different diseases with the amount of morbidity, incapacity and mortality, which they cause. The eventual idea is to relate medical research spending to the pattern of ill health in Britain. More money would be spent on illnesses that cause a social burden, because they affect a lot of people, and keep those people away from work – however minor those illnesses may seem.

Back pain is a good example. It does not kill anyone, unlike cancer and heart disease, and so for years it has been the Cinderella of medicine. But it does affect more people in Britain than any other illness, often keeping them off work, with disastrous consequences for our economy.

The Government study aims to make sure that money is being spent in all the right areas. Cancer and heart disease, for example, will continue to receive generous support because they do affect so many people. But obscure medical areas will get less money. At the moment there is a danger of excessive specialisation. Many able people are working on problems, the solutions of which would improve the health of a few, while too few people are working on those health problems which are of major importance to millions of people.

In the expectation of the Government changing this state of affairs and favouring research, however mundane, which benefits the majority, this chapter looks at three medical subjects: back pains, tooth decay and injuries. These affect us all at some time, and all have been covered regularly by *Tomorrow's World*. Many people feel these areas of medicine need more funds.

Back Pain

More otherwise fit people in Britain are disabled by back trouble than by any other illness. Every day nearly 60,000 people, on average, are away from work because of it, at a cost to the economy of £200 million a year.

It is a branch of medicine which has been neglected in the past, mainly because back trouble has been difficult to define and difficult to investigate. But now the Department of Health and Social Security, and organisations like the Arthritis and Rheumatism Council, are beginning to make back trouble a research priority.

There are new approaches in several areas. Engineering technologies are being developed to overcome uncertainties in diagnosis and treatment. At Bristol University, for example, new techniques to measure stresses in the vertebrae are leading to better advice on the correct way of protecting backs. Special 3-D X ray machines have shown that people with narrow spinal canals are more prone to suffer from backache because there is less room for their nerve bundles to escape from pressure caused by bulging discs or squeezing vertebrae. These results are helpful to surgeons operating on patients with back trouble, because they can direct attention to narrow spinal canals and pain-causing trapped nerves.

A far simpler diagnostic tool is the goniometer, developed by a doctor at Barnet General Hospital to measure flexibility of the spine. Previous attempts to assess flexibility failed because they used techniques that could not measure all of the necessary variables, such as the angle of movement of the spine from side to side, or the twisting movement. And none of them could be used to measure the flexibility of joints.

The goniometer, however, measures all these variables, assessing the angle through which the spine can move – forwards, backwards, from side to side, and in a twisting motion – as well as measuring the angle through which, for example, a leg can be lifted while the patient lies flat. It tells the doctor how much flexibility there is in each joint and assesses the amount of movement a diseased joint can tolerate.

It works on the principle of a spirit level. It has a hollow disc which is filled with a mixture of water and spirit. A pointer, pivoted at the centre, with an air bubble trapped in its tip, stays vertical when the goniometer is placed on the bent spine. The angle of the spine can then be read from a scale at the edge of the disc.

Before the goniometer was developed, assessing the angle of the spine required specialised skill and experience beyond that of the average GP or hospital registrar. But, because this device fixes the angle exactly *and* at a glance, better diagnosis will be available to back sufferers at their local surgeries.

When the measured angle agrees with the reading expected for the patient's age, it indicates to the doctor that nothing is radically wrong. The goniometer is

This refined carbohydrate and sugar diet has increased dental decay because it encourages the growth of bacteria in the mouth. Decay is the result of damage done to the protective enamel surface of the teeth by bacteria which grow in the food remains that lodge between the teeth. They digest the refined carbohydrates and sugar and form acids.

The fact that the amount of tooth decay is proportional to the quantity of sugar and refined carbohydrate in the diet was demonstrated dramatically during the Second World War, when rationing led to a steep decline in the consumption of sugary food and sweets. Dental decay dropped in parallel.

But it has taken a long time for dental research to concentrate on dietary ways of combating decay. One exciting study, however, is being done at Newcastle University Dental School, where the effects of different foods on the bacteria in the mouth are being monitored. Foods eaten at the end of a meal, or snacks between meals, have been studied first.

The effects of these different foods were assessed by removing the plaque containing bacteria from the teeth, and measuring its acidity or alkalinity before, during and after eating a food. The results have shown that the traditional advice, to end a meal with an apple, is quite wrong. Although the apple may act as a natural toothbrush, cleaning the teeth and massaging the gums, it is also acidic and contains sugar. Because of this, apples encourage the bacteria that cause tooth decay and are therefore more likely to harm the teeth than help them.

If apples will not keep the dentist away, what will? The Newcastle research shows that the ideal food with which to end a meal or eat as a snack should be non-carbohydrate and non-acidic. The present favourites are cheese and salted peanuts. Both have a bonus. They are good at stimulating the flow of saliva, which helps to combat decay by washing away the harmful bacteria.

Our table of foods, tested during this research, in order of merit, show some surprising results. It runs as follows: cheese, peanuts, crisps, bread, ice-lollies, apple, ice cream, chocolate, chewing gum, biscuits, chocolate biscuits, orange drinks, sugared coffee, and boiled sweets. The apple, far from being the dentist's delight, is down below ice-lollies because of its harmful acidity. Chewing gum, which many people believe is good for cleaning teeth, contains too much sugar to be safe, and those parents who give their children orange drinks should notice how far down on the list this appears. Many people today end their meals with coffee, often including sugar. They would be wise either to leave out the sugar, or replace coffee by cheese.

There is obviously a lot that we can do ourselves to help prevent decay. We should eat less sugar and other carbohydrates, and more wholemeal flour and raw vegetables, as well as ending meals with something non-acidic and non-carbohydrate, like cheese.

We should also make sure that we use toothpaste with fluoride, and we should replace our toothbrushes at least twice a year. This is obviously not done, despite dentists' warnings, because the number of toothbrushes bought annually is lower than the number of toothbrush users in the UK. Using plaque-disclosing tablets is also a useful habit.

Another quite different approach to preventing decay is vaccination. In small-scale tests on children, a new vaccine cut the rate at which tooth cavities form by 85 per cent, by stopping the formation of plaque on the teeth.

The theory behind the vaccine is simple. Natural bacteria in the mouth are normally harmless, but there is one particular bacterium which forms the plaque by mixing with processed sugar and forming dextrans. This sticky material releases enzymes which form the acids. The vaccine breaks the process, by inducing the production of antibodies which stop the mixing of sugar and bacteria.

In clinical tests the vaccine is given once a year, starting before the first teeth form. It is injected into the lining of the mouth, so that the antibodies form in the saliva. Future problems seem to lie, not with the vaccine, but with the National Health Service's policy towards preventative dentistry. As I explained earlier, preventative treatment like vaccination is not paid for by the NHS, which obviously limits its availability and effectiveness.

While we concentrate on treating dental disease rather than preventing it, we can continue to expect dramatic changes in dental technology, like the lasers, ultrasonics, air abrasive 'drills' and implants that have already arrived in dental surgeries.

Ten years ago the patient could expect to be treated by a dentist using a Christmas tree-type of dental unit, which held the drills, lights, air, X ray, instruments and water sprays. Today's modern dental units have lost most of these appendages, and look far more like an efficient combination of a trolley and chest of drawers. The patient is likely to lie on an upholstered couch, which can move into almost any position, or even on a waterbed. The dentist is also more likely to work in comfort, in some kind of seated position.

Probably the biggest advance in equipment he uses has been the introduction of the air-turbine drill. Rotating at up to 600,000 rpm and using tungsten carbide or diamond drills, it strokes away the tooth with great precision, allowing complicated work on crowns, bridges and other advanced restorations.

Ultrasonics are used for scaling teeth and removing stains, while quartz halogen lights are common for general illumination. For more precise lighting in the mouth, the fibre optic principle of sending light along several feet of cable is used.

A different type of light, ultra-violet, is used as a catalyst, activating the setting mechanism of new filling materials like plastic resins. There have been two recent advances with these resins. First, the simple resins have been loaded with inorganic materials like quartz, which improve their ability to restore and build up teeth. The second advance has been to etch teeth, creating microscopic crevices, before adding the filling material. This technique has made it much easier for dentists to treat children's broken front teeth, for example, without causing pain.

A new adhesive filling also reduces the pain normally associated with dental treatment. Called glass ionomer cement, it eliminates the need for extensive drilling into healthy teeth to provide the undercut anchorage needed for conventional dental cements. The hole in the tooth now only has to be widened enough to remove the decay before the adhesive filling is packed into the unlined cavity. As it sets, strong ionic bonds which hold the filling firmly in place are formed between the cement and the tooth surface. It is expected to be used by dentists treating children or repairing the grooves that often appear as an adult's gums recede, exposing the softer part of the tooth.

In Britain every year 10 million permanent teeth are extracted and replaced by dentures, at a cost of £30 million. So any method of saving teeth is obviously good news to the dental profession. Many adults' teeth are extracted because the pulp cavity loses its normal nerve and blood supply, after disease or damage during an accident. The tooth may blacken, it can cause a lot of pain, and once it has been pronounced dead, it is extracted.

But there is a problem in producing confirmation that a tooth is dead, without active nerves and blood vessels in the pulp cavity. Teeth can be tested to check the nerve supply, either by using a small electric current, or by applying heat, and then a cold ethyl chloride pad, onto the tooth surface. If the patient feels slight pain, rather like that experienced when eating cold ice cream, it means the nerve is very much alive.

Teeth can, however, still be alive, with a healthy blood supply, without nerves. The pulp cavity often loses its nerve and blood supply temporarily, after accidents, dental operations or disease. And the blood supply always returns first, often months before the nerve. Therefore the dental profession has been looking for a test to confirm that blood vessels exist, because it would be the first sign of a 'dead' tooth coming back to life.

Recently such a test was developed, which involves using two probes to record the temperature of the suspect tooth on its surface enamel and the mouth temperature next to the tooth being tested. The suspect tooth is then heated with a rubber cup, revolving at a set speed, using a precise pressure, for a certain length of time. Then the time taken for the tooth to cool to its original temperature is checked.

If the tooth is alive, with a healthy blood supply, the heat increases circulation, so the pulp heats up. But a dead tooth, with no blood supply to affect it, is only influenced by the atmosphere of the mouth, and therefore returns to its original low temperature far more quickly than a living tooth. This new test could save some of the 10 million teeth now lost every year, and will mean that patients go to their dentist to have their temperature taken too.

Injury

Another medical area which, like back trouble and dental disease, affects the majority of us, is the treatment of wounds and other injuries. Whether it is a simple cut, a burn, or the result of major surgery, today's wound is normally treated in exactly the same way as it would have been years ago, with plasters, bandages and other traditional dressings.

But these cotton-based dressings can be far from ideal because they harbour germs, they often dry out, causing the wound to form a bad scar, and they can also stick to the damaged part of the skin, causing extra problems when they are removed.

For these reasons, research has been going on for some time to find the perfect wound dressing. Doctors wanted a dressing which would create the best possible micro-climate for rapid healing, with four particular properties. The wound surface should always be moist; but not wet. Oxygen and carbon dioxide should be able to diffuse freely around the wound surface. Excess fluid should be able to evaporate through the dressing or be absorbed by it, and any new tissues that form should not become incorporated into the dressing.

After trying many different materials, a doctor at the Institute of Orthopaedics, helped by research at the Medical Research Council Burns Unit, discovered a completely different way to heal. After five years of research, he decided the ideal dressing for wounds should be made of foam.

This came as a surprise, even to the manufacturers of polyurethane foam, who have seen this material used in dozens of different ways since it was first invented thirty years ago. But, of the thousands of tons a year that we use in Britain, most of it is hidden inside chairs, mattresses and cushions, as well as under our carpets. This latest, medical use, however, brings foam right out into the open.

Foam is a simple and economical dressing, partly because foam itself is cheap, and partly because it is used on its own without backing materials. During manufacture the foam is treated with heat on one side to produce a smooth surface. This smooth side is put directly on to the raw wound, where it gently absorbs excess blood and other liquids seeping from the damaged tissue.

It takes up the bulk of any liquid without drying out, which means that the wound always has a thin film of moisture next to it, a factor important for healing.

The other advantages of a foam dressing become clear when you look at the way it helps damaged skin to heal. When you examine a section through the skin, it looks like a tiered cake, with a thin layer called the epidermis on top, and a thicker layer of connective tissue, called the dermis, underneath.

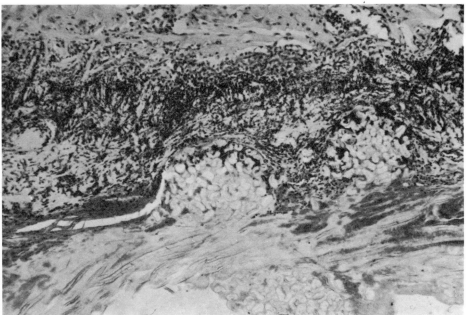

Greatly enlarged section of two identical wounds. That in the top picture healed perfectly under foam. In the lower picture a scar developed under a traditional cotton bandage because the epidermal cells did not multiply

When the skin is damaged, it is the epidermis which comes to the rescue. Epidermal cells multiply and move under the wound to form a new, outer skin. But the speed at which these epidermal cells move, and therefore the wound heals, depends on the amount of oxygen reaching the area of damaged tissue.

Foam has a great advantage over cotton-based dressings because oxygen passes easily through the intercommunicating honeycomb of foam to reach the moving cells in the epidermis. At the same time, the foam slowly absorbs blood from the wound, plus important white cells in the blood. These lurk in the honeycomb of foam ready for the germs which always collect around wounds. They engulf any bacteria as they reach the foam, so that the chances of infection are greatly reduced.

Also, as the new outer skin forms from the epidermal cells, the smooth surface of the foam next to the wound does not become incorporated into the new tissues. There are therefore fewer problems when the dressing is removed.

Foam speeds up the healing process by a factor of three because of the combination of its special properties which add up to an ideal environment for healing: the foam's one-way absorbency, which keeps the wound moist, but not wet; its ability to allow oxygen to get through to the wound; and the smooth surface which is put directly on to the raw wound and does not become incorporated with new tissue.

The consequence of this environment can be seen in a microscope preparation of two identical wounds. The first, which healed under foam, shows the epidermis having grown under the wound, quickly and cleanly. The second wound had a traditional cotton dressing. The epidermis did not multiply and move under the wound as fast as it should have done. The result is easy to see in the picture – a nasty scar.

This foam is being used in all sizes. There are large dressings for hospital skin grafting, major surgery and burns, while smaller dressing-strip versions bring the same quick-healing advantages to simple cuts and grazes.

Large sheets of a slightly different foam are also being used as an emergency burns dressing by first-aiders. After a bad burns accident, the patient's whole body can be wrapped in a foam cocoon, which is thick enough to give protection and will not stick to the wounds like traditional bandages. Ambulance services and fire brigades use this foam for initial first aid and protection during the journey to hospital. It is applied dry, completely sterile, straight from its wrapping, which speeds up the whole process of dealing with bad burns, when every second counts.

Broken Bones and Damaged Ligaments

Another common injury is a broken bone, and the most exciting advance in this medical field is the use of electromagnetic waves to heal fractures. The technique has been used in New York to repair broken bones in adults and to treat a condition in children known as congenital pseudoarthrosis of the tibia. This involves breaks in one of the lower leg bones, and, until now, because the nervous tissue is defective, healing has been extremely difficult. But this new method has proved successful for 85 per cent of these children, and for adults with fractures that were resistant to healing.

The medical world is so excited about this technique, which cures without the need for a doctor to 'invade' the body, that it is now being used in twenty other institutions around the world. It may not be long before this completely non-invasive treatment is available to the majority of patients with problems over fractures.

It works by sandwiching the fracture, still in its plaster cast, between two coils of wire. Current is pulsed through the coils to create an electromagnetic field. This field penetrates the bone and soft tissues, inducing a pulsing current. Healthy cells ignore the signals, but in areas of chronic bone injury the cells respond.

It is a major advance on research into bone metabolism twenty years ago, when bone growth was induced around an electrode. But internal electrodes were found to be dangerous, as well as involving undesirable surgery.

This new electromagnetic treatment has halved the healing time of run-of-the-mill fractures. Patients can wear the equipment at home for the necessary fourteen hours on average each day, which reduces the costs of keeping them in hospital. Soon this equipment will also be battery powered. The method may have implications far beyond the treatment of fractures. It allows cell membranes to be manipulated, so that the way in which they allow substances to pass through them can be altered.

Tendons and Ligaments

Sportsmen and old people with degenerative diseases, like rheumatoid arthritis, often suffer from damaged tendons and ligaments. Surgeons have tried to help severe cases since the early 1900s with silk and nylon tendon and ligament replacements. More recently, braided stainless steel and Teflon-coated polythene have also been tried.

All of them are string-like in appearance, and like string they have a disadvantage. They snap – often within three or four years of the operation.

In the search for something stronger, surgeons at the University Hospital of

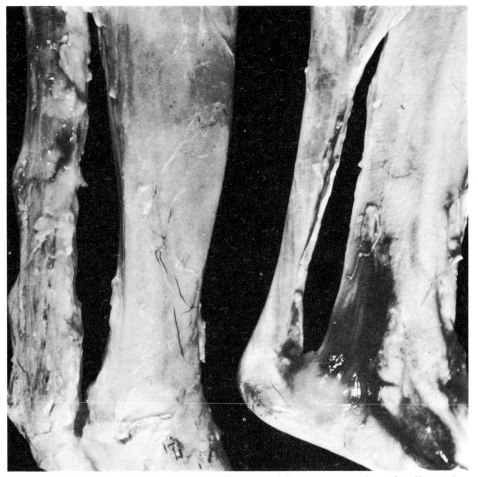

On the left of this picture you can see the new collagen which has grown around a carbon fibre tendon replacement after six months. It is beginning to look like the undamaged tendon on the right

Wales tried carbon fibre, the tough, light material that is used for making fishing rods, ski poles, ice-hockey sticks, helicopter blades, Concorde brake linings and many other things that have to take a lot of strain.

At first the normal, rigid carbon fibre was used. But when the surgeons saw the raw material, which is used in the factory to produce rigid carbon fibre, they realised that it was far more suitable for the job. So now damaged tendons and ligaments are replaced by soft carbon fibre, twisted for extra strength.

The operation on a damaged knee is short and relatively simple. The surgeon first drills two holes in both the femur and tibia, the bones which join at the knee. Then the twisted carbon fibre is threaded through the holes to strengthen the knee and do the job of the damaged ligament.

Once the carbon fibre is in the body, its magic really starts. The surgeons doing this work do not understand why, but the carbon fibre replacement actually encourages the production of natural tissue. Collagen, the dense connective tissue that makes up ligaments and tendons, grows around the carbon fibre within six to eight weeks of the operation. The other great advantage of this method is that in time the carbon breaks down and is absorbed harmlessly into the body.

The operations have been so successful that rugby players have been able to go back to the game with carbon fibre replacements in their knees and ankles. The work is at too early a stage, however, for doctors to raise hope for patients with problems in their fingers and hands.

But there is a pleasing conclusion to this medical story. Carbon fibre is also helping the animal most prone to leg injuries, the racehorse, which was not designed to gallop at 35 mph throughout a long career. Not surprisingly, the majority of racehorses suffer at some time from tendon trouble. It is an expensive business when a top stud-value Derby winner is worth £2 to £3 million.

Owners will do almost anything to get their horses back to work after tendon trouble. One common, though controversial, treatment is called firing. With the leg anaesthetised, a hot iron is drawn across the skin, leaving scars, which, some people believe, support the damaged leg and encourage the tendon to heal. The critics, however, say that firing is both painful and useless.

But now there is a medically-sound solution. Carbon fibre replacements are being used for racehorses with damaged tendons, giving them the chance to race again. It is good to hear of work done on humans proving useful for animals.

Medicine of the Future

This chapter has looked at just three of many medical subjects, which have been covered extensively on *Tomorrow's World* because they affect most of our lives. The individual items have been developments of the 1970s, but we are also fascinated by the future and wonder what the remaining two decades of this century will bring.

The three subjects already discussed will not escape change. It is fairly safe for me to predict that by the year 2000 or not long after, there will be a need for far fewer dentists. New knowledge of diets will have prevented tooth decay. And for patients too old to benefit from future dietary changes, tooth implants with carbon roots may take the place of dentures.

Teaching programmes, concentrating on the correct ways to sit, lift, push and pull could reduce back trouble significantly, and injuries will be less of a problem. We already have a method of replacing ligament and tendons with a material which encourages the regeneration of collagen. In the future I expect that we shall also see the regeneration of bone, and possibly, all vital organs.

While bone replacements will be considered standard surgery, and with rejection problems overcome, we may see successful implants of artificial bladders, kidneys, pancreases, livers and hearts. Organ transplants to correct certain genetic deficiencies are expected to be routine within the next five years. There is a general optimism about the future. Specialists in most fields of medicine now claim that they are beginning to understand their subject. In the past we have accepted a kind of trial-and-error therapy, because doctors were bewildered by the underlying mechanisms of disease.

But now we are entering a new era, when these diseases are being treated as problems that can be solved. Medicine is becoming less and less an Art, and more and more a Science. A science which depends on new technology.

I believe that automation and electronic processing will be routine in future laboratory diagnosis, particularly in microbiology, cytology, haematology and chemistry. Completely computerised general health care centres will provide mass medical care, while home-based equipment will allow twentieth-century-style 'house calls' by telemetering clinical findings to a base staffed by doctors who will make the diagnosis and telephone through the prescription.

America already has telemedicine. A doctor in one part of the country examines a patient in another, using television techniques. Electronic controls allow the doctor to adjust, remotely, the television camera at the patient's end. He can make it zoom in and out, focus, pan and tilt. The doctor can then diagnose from a distance, taking into consideration heart and breathing rates, X rays, electrocardiograms, electroencephalograms, microscopic examinations of blood samples and tissue, and fibre-optic views of the body.

But automation of this kind will not mean the end of the doctor. I hope that he will be left with more time to listen and look after his patients, in a more comprehensive, caring way.

There are certain diseases which everyone hopes will be conquered by the year 2000. There has already been remarkable progress with what must be today's most frightening disease, cancer. The outlook for children with leukaemia is brighter. As *Tomorrow's World* has shown, anti-cancer drug therapy is offering hope for women with breast cancer. And there have been advances in the treatments for Wilm's tumour of the kidney in children, choriocarcinoma in women, Burkitt's lymphoma and Hodgkin's disease.

In the next twenty years I hope we will see specific chemotherapy or immunotherapy for many types of malignancies. Priority will probably be given to identifying carcinogens in the environment, and the environment will be carefully screened for these carcinogens by bacterial and mammalian cell systems.

Heart disease, which is responsible for the greatest toll on human life, may also be reduced in medicine because of a better understanding of blood pressure control, a massive lowering of the circulating blood lipid concentration, and the discovery of all causes of arteriosclerosis, which is said to be the basis for 95 per cent of heart ailments.

The treatment of mental illness should also improve, with drugs to treat schizophrenia and depression. There will be more development in identifying chemical bases for psychiatric disorders. The combination of anticipatory guidance to help patients cope with specific stressful circumstances, plus drugs to control personality, should make insanity and suicide much less common by 2000.

I also hope that we shall see worldwide immunisation against infectious diseases, the treatment and possibly even the cure for crippling diseases like multiple sclerosis, muscular dystrophy and rheumatoid arthritis, and the non-narcotic treatment of pain.

Scientists may even make old age less harrowing. They may even find a cure for senile dementia, which is now the largest health problem in old people. By the year 2000, people of 90 plus should be healthier and more productive.

Other goals would be to eliminate the hazards of being born prematurely, solve the problem of the common cold, reduce stress so that men live as long as women, and to have target-directed drugs. It has already been predicted by some doctors that similar techniques to those used in the drug industry will produce slow-release food capsules, which will allow us to go for weeks without eating.

We must also have faith that society will be able to cope with the moral and social issues arising in the future from these scientific discoveries and changes. It is going to be a *very* different world tomorrow.

Michael Rodd

Travelling Hopefully

It goes a little against the grain to have to start a chapter on the transport which will carry us into the eighties with an admission that most of us in the next ten years will be travelling in vehicles not very different from those of today. Certainly, if the world's oil runs out sooner than predicted we may find fewer of us can afford to run private cars and may have to travel by bus or train – but it will be, for most of us, a bus or a train and not a tracked hovercraft powered by linear motor, computer-controlled monorail, or any other futuristic system of public transport.

And yet the mid-seventies have been a time of real achievement in transport, coupled with a general realisation that the development of conventional technologies produces more rewards faster than the expensive business of starting a new technology from first principles. Concorde is in service, but the world's airlines are buying the younger and technically much less sophisticated European airbus, the A300B. The Japanese Shinkansen high-speed trains have been moving at 130 mph on their special tracks for ten years but elsewhere in the world the trend is to develop trains like our own high-speed diesel, which will travel at that speed on conventional lines amongst ordinary services. In America, RCA and General Motors may both have demonstrated automatic road-vehicle steering in the late fifties, but to date, the only practical application I have discovered is the automatic baggage carts system at airports like Charles de Gaulle, Paris. There, the driver-less vehicles steer themselves by following a buried wire the short distance from aircraft to baggage hall.

Both Germany's Mercedes-Benz and Britain's Transport and Road Research Laboratory at Crowthorne in Berkshire have recently tested an automatic driver-less bus using the same technique – but neither they, nor RCA, General Motors, or anyone else, have yet seriously challenged the vehicles with human drivers seen in high streets all over the world. So where is this achievement I was talking about earlier? The answer is in the developments within our conventional transport technologies.

Battery Prospects

Let us begin our detailed look at the prospects for transport with a British story in which one major developer believes in pursuing new technology whilst its major rival is capitalising wholeheartedly on what can be achieved now.

The prospect of quiet, pollution-free cars and buses receiving their energy from never-ending wave-generated electricity is a beguiling one. Yet it is a prospect almost as distant today as it was when we reported on a battery-powered prototype in 1974.

At that time, it was believed that the battery itself was holding up the progress of the electric vehicle. To get enough power to drive a vehicle any distance at a useful speed meant having so many heavy lead-acid batteries on board that there was not enough room for a useful payload. The breakthrough was going to be the development of a sodium-sulphur battery which would give five times the energy for the same weight of lead-acid battery. In the new battery, the electrolyte is solid beta-alumina, and the two electrodes, one sodium, the other sulphur, are molten.

Today, only one British firm, Chloride, as part of a consortium with the Electricity Council, is pursuing the sodium-sulphur battery. The firm admits it has sunk over a million pounds into the project, but it still talks of, 'If we succeed', rather than, 'When we succeed'. The problems are substantial. The battery has to be used at 350° Celsius. At such a high temperature sodium is a particularly tricky substance to handle and highly corrosive. But the commercial prize for success will be enormous, and because the Japanese, the Germans and the Americans are also in the race, the British sodium-sulphur contender is prepared to give nothing away, save that the development is a tough one.

And yet the electric-vehicle story has not stood still whilst waiting for the big step forward that the new battery will bring. Both Chloride and another major manufacturer, Lucas, are busy now, extending our existing technology.

In a conventional lead-acid battery the acid is the fluid electrolyte through which charged ions pass from a lead plate to a lead dioxide plate. This was the design which in 1974 was holding up the development of the electric vehicle. Since then lead-acid has been further researched. A much lighter polypropylene outer casing has reduced its weight. A re-design of the inside allows more of the active materials to interact with one another, and the vehicles themselves have been rethought to eliminate unnecessary losses of power.

One engineer in a car plant is reported to have been surprised that an electric vehicle builder thought a 15 per cent loss of power between engine and drive wheels was unacceptable. 'We calculated,' the electric vehicle builder told me, 'that to cope with that 15 per cent loss, we would have to include at least 20 per cent extra batteries, because the more weight in extra batteries there was to carry around, the more batteries we would need, and so on.'

By redesigning the whole power transmission system, it has proved possible to reduce those losses to just 1 per cent under some running conditions. Out go both the transmission shaft running the length of the vehicle and the differential connecting it to the back axle. In their stead is a special back axle assembly and a chassis-mounted electric motor close alongside the wheels.

It is interesting to observe that Chloride, which is developing sodium-sulphur batteries, still uses a largely conventional transmission and differential design. It is Lucas, relying on getting the most out of existing battery technology, which has produced the unconventional power transmission system.

And what sort of vehicle are they both manufacturing? The silent bus, the city taxi, or the second family car? None of these in any quantity. Attention is being focused on the urban delivery van.

Research into how vehicles are used revealed that a large proportion of small delivery vans – with less than 2 tonnes payload – travel almost fixed mileages every day over a predetermined route. No other class of road vehicle has such a predictable work pattern. Lucas decided to try and produce, by 1980, a 1-tonne van with a 70 miles range and a top speed of 55 mph. Chloride went for a larger van with, initially, a lower range and top speed.

Could electric vehicles be developed which had enough acceleration and speed to be capable of fitting into a fast-moving city traffic and which could carry useful payloads? It was the first prototype of just such a vehicle which we presented in

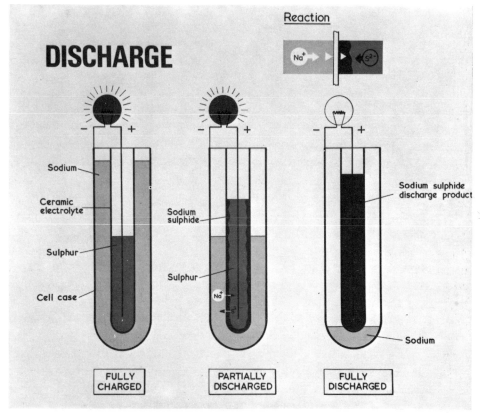

The sodium ions and sulphur ions change places producing the electric current

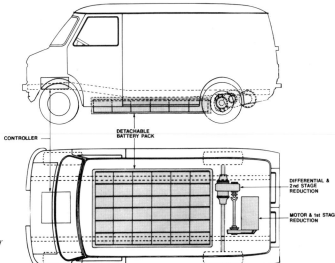

CONTROLLER

DETACHABLE
BATTERY PACK

DIFFERENTIAL &
2nd STAGE
REDUCTION

MOTOR & 1st STAGE
REDUCTION

*The battery pack and power
unit of the Lucas van*

our studio in 1974. Since then, it has become clear that quiet, pollution-free running would not be the only advantage to be gained from electric vehicles. They do also promise to save money. Maintenance costs are a fraction of those of their conventional counterparts. Electric vehicles have far fewer moving parts, suffer less vibration, and therefore less stress on the body. They also can manage without a clutch or gearbox. These, in a conventional vehicle, receive a tremendous hammering in any city centre where one is frequently stopping and starting. But the benefits of cheaper maintenance are at the moment outweighed somewhat by the expense of buying an electric vehicle. Today there are probably less than 100 urban electric delivery vehicles, excluding our much slower milk floats, in operation in Britain. As more operators become convinced of their value, more will be produced.

But until production lines can be set up, electric vehicles will have to be one offs, rolled off production lines at conventional vehicle plants, to have their drive systems installed elsewhere. That means that even without the batteries, which might cost a thousand pounds or more a set, an electric vehicle will cost at least one and a half times its conventional petrol or diesel-driven rival. It will be like this until mass production savings have effect.

But Lucas reckons that if its small van could be used for more than 5000 miles in a year it could prove an economic proposition. Taking into account the reduced maintenance costs, Chloride's larger electric vehicles start to save money, at today's prices, in their seventh year of operation. But they may have a life of fifteen years, and it has already shown that they can carry a $1\frac{3}{4}$-tonne payload, travel 40 miles on one charge and accelerate to 30 mph from standstill in twelve seconds.

So why no similar development of the electric bus, the city taxi and the family runabout car? The average second family car does not do a large enough mileage to justify the expense of an electric vehicle, and the main family car is called on to travel further than the 40-mile limit. As battery charging takes several hours, filling up at the motorway services would be impractical. Even replacement batteries may not provide the answer. Few garages could afford a stock of replacement units, with each one costing well into four figures. Any taxi with an uncertain daily mileage suffers from the same limitation, and the annual production of buses that could be electrically powered, single deckers principally for short routes, is so small as to be commercially of little interest.

Even the imaginative but experimental white cars of Amsterdam have not yet been used elsewhere. As regular viewers will remember, these are strange, cylindrical, battery-powered vehicles which can be hired for self-drive travel between city centre charging points. But we can certainly look forward to more city deliveries being made electrically, with the promise of a more complete electric revolution on our roads when, or rather, if, the sodium-sulphur breakthrough is made.

Moving by Gas

Whilst the electric vehicle battles to become established in its logical place in the market, the gas-powered vehicle is already ahead of it. British road vehicles now consume about 10,000 tons of liquefied petroleum gas (LPG), every year. Even though our consumption of LPG is growing, it is still minuscule compared to that of Holland; there, vehicles use 800,000 tons a year, an amount that makes the Dutch the world leaders in this field.

It is likely to receive a substantial boost when manufacturers announce the first totally LPG powered vehicle in Britain. Ford, Rolls-Royce and Volvo, for instance, are actively developing LPG engines.

Up to now, this gas has been consumed by cars and trucks which can run on both conventional petrol and LPG. Because an engine which can cope with both conventional fuel and LPG has to be a compromise, it is argued that it does not make the most of either fuel. Yet if a car is going to travel more than 20,000 miles a year, it will be cheaper today if it can do most of that mileage on gas, not petrol.

It works like this. The tax on LPG for road vehicles is half the tax imposed on petrol. This was a measure adopted in 1971 to reflect the amount of energy contained in a gallon of LPG and in a gallon of petrol. The cost of converting an all-petrol engine to gas/petrol is now around £200. So the advantage of the lower price of a gallon of LPG is offset by first a lower mileage per gallon when running on gas and second the cost of the modification. Any mileage under 20,000 a year is unlikely to be enough to swing the balance in favour of LPG. But commercial drivers do much more than 20,000 miles a year, and it is becoming more common to find long-distance motorists filling up at the growing number of motorway service areas and garages which also offer LPG facilities.

A car that has been converted for gas carries a small vapourising unit which draws warm water or exhaust gases from the engine and liquid gas from the storage tank. Once the LPG is warm and vaporised, a vacuum valve connected to the engine's inlet manifold opens and the petroleum-based vapour is drawn into a slightly modified carburettor. The vacuum valve ensures that, should the engine stall, the neat gas supply is cut off altogether.

In the engine, the gas and air mixture fire in the same way as the petrol and air mixture normally does. There is evidence that the quantities of pollutants like unburnt hydrocarbons, carbon monoxide and carbon dioxide are reduced when vehicles burn gas rather than petrol. A system has even been developed for heavy lorries, enabling them to mix LPG and diesel fuel, and so go uphill without their engines producing clouds of black smoke. Though we demonstrated the principle working on *Tomorrow's World*, it has yet to be taken up seriously.

If LPG has a weakness, it stems from its close association with oil. First, the real profits in the oil industry come from selling large quantities of oil, not from selling the comparatively small quantities of LPG which can be produced from the same batch of crude oil. Consequently, oil platform operators in the North Sea often flare off the LPG as an unimportant by-product, rather than embark on the more time-consuming exercise of getting both LPG and crude oil ashore and refining the whole lot.

The UK government has insisted that this potentially valuable energy source is no longer wasted. An estimated extra three million tons of LPG a year is to become available to Britain. This will not only affect the price of LPG, it will also provide enough stocks to enable the fuel to fight for a much larger share of the market.

The second weakness of LPG is that it is petroleum gas, and when the oil runs out, so will the LPG. So it is not going to solve our long-term needs. However, in the next ten or twenty years I would not be surprised if more of us find ourselves riding in cars that really do run on gas.

But are there no revolutionary answers to our mundane transport problems?

Rent-a-Monorail

On a test ground in West Germany, scientists have built an experimental monorail taxi service. The principle is straightforward. At stations on some future urban route it will be possible to pick up a small electrically-powered vehicle, programme the central computer with a destination, be billed for that trip automatically, and pay the fare into a slot inside the vehicle.

The West German scheme worked by transmitting a digital code which is picked up by trackside detectors as the vehicle moves forward. This changes points to take the vehicle the easiest and least congested route to its destination. But brilliant though the concept may be, there have been no plans to install it in any city and it remains purely a test project.

Metro on Tyneside

No such problems exist in North East England. The Tyne and Wear Passenger Transport Executive is about to open one of Britain's biggest railway projects since the last war. The 34 miles of the Tyneside Metro owe almost as much to George Stephenson as they do to the sophistication of today's engineers. At least one of the viaducts the line will use was built in 1839. Everything in the Metro has been done at least once somewhere before. This approach is to ensure that the 900,000 Tynesiders get a transport system which works.

The Metro got going very quickly. First detailed in 1971, it was in the process of being built four years later. Approval came from the Government, it is said, largely because the plan makes substantial use of an existing rail system. This was built before the turn of the century and has been allowed to run down. Recently, it has cost £1½ million a year to subsidise. Develop it, or shut it and find some other way of getting the Geordies about, were the alternatives open to the local transport executive. The Government agreed to foot 75 per cent of the cost of converting the old local railways to rapid transit and, in 1975, *Tomorrow's World* was there when the first sod was ceremonially cut.

The major new works, tunnelling under Newcastle and Gateshead and constructing a new crossing of the Tyne, are only a small part of the whole system. Much of the line will be the original nineteenth-century commuter routes which linked Newcastle with the coast at Tynemouth, Whitley Bay and South Shields. Overhead electric cables will provide 1500v direct current to light-weight rail vehicles, 'Super Trams' they have been nicknamed, running on standard gauge railway track. The twin cars will carry 84 seats and have room for 188 standing passengers. They will average 25 mph, and have a top speed of 50.

The low weight of the metro cars will greatly reduce maintenance costs on track originally laid down to take full-scale trains. Many of the originally magnificent stations will be reduced to platforms and shelters. Station staff will move from halt to halt rather than having stations manned all the time. The aim of the Metro is to reduce the unnecessary and expensive frills of Victorian railway engineering, whilst capitalising on the sound infrastructure to meet today's needs.

Where rerouting has been undertaken, it is to bring the system closer to the people who will use it. The new tunnel under Newcastle will bring the Metro right into the centre of the city, just 15 metres under John Dobson's famous

Grainger Street, where the original Victorian engineers did not try to build their railway. In South Shields, a new stretch of the Metro will take it through the town, not along the now quiet riverside which was the scene of all the action seventy or eighty years ago.

The line will bring back into use a branch from South Gosforth to Kenton which last carried passengers before the 1939–45 war. As demand rises, it may become possible to extend this branch to Newcastle Airport and beyond to the prosperous commuter area of Ponteland.

The Metro has not had an easy time of it. Railwaymen did not like the idea at first, because they feared they would lose their jobs to the busmen. Then inflation hit the project. An initial construction figure of under £70 million now stands at nearer £200 million – and the local authority has had to accept that the 75 per cent government grant has a ceiling. Any increase above that ceiling, £161 million at 1975 prices, must be borne locally. But the Metro goes ahead.

It has been under attack more recently because all major road improvements have had to be shelved to make money available for the project. But it has already had hidden benefits. Thanks, partially at least, to the experience gained on Tyneside, at least one major British contractor is now working on the Hong Kong Metro and is actively pursuing a chunk of the valuable contract to build Mexico City's proposed rapid transit.

Metro's advocates are convinced that their decision to avoid trail-blazing technology and to keep it simple, reliable and comfortable will pay off, both in terms of overseas contracts for British industry and in increasing passenger totals and revenues on Tyneside.

Let me now spend just a short time looking at some American rapid transit systems serving similar purposes to Tyneside's Metro.

Metro Stateside

Washington DC is in the throes of launching its own Metro, 198 miles long. The trains average 35 mph with five minutes between peak-hour services, and have a top speed of 75 mph. Latest estimates put the construction's cost at around 5 billion dollars. The project has been talked about seriously since March 1954.

In Atlanta, Georgia, a similar system christened MARTA is being installed; it includes 50 miles of new rapid rail transit. The cost of setting up a combined train and bus operation will be at least 1300 million dollars. But MARTA is a newer idea than Washington's Metro. The first proposals for MARTA were drawn up as recently as 1960.

But San Francisco has perhaps the most visionary of all rapid transit systems; it is called BART. It was first discussed back in 1947. Connecting the cities of

Oakland and San Francisco via an underwater tube, and with lines linking the outer suburbs of each city into the system, the Bay Area Rapid Transit System is a little over 70 miles long. Its entirely automatic trains can operate at just two-minute intervals, though under normal conditions, six minutes separate each train. They have a top speed of 80 and average 39 mph. The cost of BART and its disappointingly small numbers of passengers have been prime considerations in the mind of President Carter, who has called for a reappraisal of US urban transport policy. His administration now advocates 'light rail' systems. Could he be thinking of something like the Tyneside Metro? Certainly the British system, even at £200 million plus, starts to look really quite good value.

Flying for More of Us

In aviation circles, the mid-seventies will be remembered as the time of Concorde. Whether the mighty aircraft is destined to become another relic like the Spruce Goose of Howard Hughes or the Bristol Brabazon, remains to be seen, but she will always be remembered as the first airliner to carry her passengers through Mach 2, twice the speed of sound. I would have said the first to carry them through Mach 1, but I remember a conversation with a French aircraft designer who told me that on occasions conventional aircraft nudge the sound barrier. But that is a statistical niggle. As a technological achievement, Concorde is in a quite different league from any conventional airliner.

But, when you and I take our next airtrip it is unlikely to be in Concorde. In the real world it is not simply technological achievements which produce success. When we travel by air, we are far more likely to travel in another European aircraft of the 1970s, an aircraft much less dramatic than Concorde, but arguably much more in tune with the demands of airlines and air travellers.

The A300B, the airbus, is a supreme example of technological achievement within conventional constraints. True, it has yet to secure enough orders to break even, but it has, for some time, been the best-selling, wide-bodied aircraft in the world and looks set to make substantial inroads into the American market.

Conceived later than Concorde, the airbus was designed to fill a gap in the range of conventional passenger aircraft. Airlines tended to use aircraft like the British 1–11, or the American DC9 or 737 for their short-range operations. Medium-range flights were handled by a larger version of the DC9, by Tridents or by 727s capable of carrying over 100 passengers. Long-range flights were usually taken care of by 707s or DC8s with the British VC10 appearing in all too few airline fleets. The range was a tidy one; the smaller aircraft carried 80 or so passengers, and the largest around 160. But when Boeing's plans for the Jumbo became known, airlines had to rethink their policies. The prospect of a long-haul 350-seater left a large gap in the airlines range for the early 1970s.

A consortium of Europeans – the West Germans, the French and the Spanish – set up a production line at Toulouse in Southern France to try and fill that gap. Its aim was to build a conventional aircraft based on known technology which could carry large numbers of passengers over short- and medium-haul distances very economically.

Britain could have made that consortium a quartet, but chose instead to finance the development of Rolls-Royce's RB 211 engine for Lockheed's Tristar airline which was being built in America. As a result, it is American, and not Rolls-Royce, engines which power the airbus, and Britain has had a limited role in designing the aircraft. Happily, the Hawker Siddeley company, now part of British Aerospace, did get involved with the airbus, designing and building the wings, thereby maintaining some British interest in the project.

The wings are an interesting aspect of the airbus, because they illustrate how the aircraft has taken conventional design to the limits. They are a mass of slats and flaps, which greatly extend the wing area during take-off and landing, and which fold away in flight leaving a wing which distributes lift more evenly than the wings of older aircraft, delaying the formation of shock waves and greatly reducing drag.

This design improvement is not used to make the aircraft fly faster than its rivals, but to fly at the same speed with wings which are much less swept back. The straight wings of the airbus require less reinforcing structure than the wings of other conventional jets – saving around a ton of metal, a ton which can be added to the aircraft's payload. That greatly appeals to airlines who would far rather lift profitable passengers and freight than unneccessarily heavy wing reinforcement.

A later development called the Kruger flap is a tiny, but aerodynamically important, extra spoiler which cleans up the wing line when the front flaps are extended. This one addition to the take-off equipment gives the airbus an extra seven tons of payload capacity when operating out of high-altitude airfields where the air is thin and lift harder to achieve.

The airbus is the only wide-bodied airliner to use just two engines. Jumbo 747s carry four, and the other American wide-bodied craft, the Tristar and the DC10, both carry three. Using only two engines means the designers in Toulouse have produced an aircraft at least 15 per cent more efficient over its designed operating range than the Tristar or the DC10, its nearest competitors.

The components are manufactured in specialist factories all over Western Europe. When I visited the production line in Toulouse I watched the company's super Guppy arrive. This is an American transport plane based on the Jumbo of the 1950s, the Boeing Stratocruiser, which has an enormous bulge in its midriff. On this journey, it was carrying wing sections which had started their journey in Chester, England, flown for component-fitting to Bremen, West Germany, before coming on to Southern France. 'It will be fuselages from Hamburg tomorrow,' said the unloading supervisor, anxious to practise his English, 'or maybe tail planes from Madrid.'

But what is perhaps most significant about the A300B, is that it can be developed further. It can carry four engines and be made longer, it can be made shorter and therefore will need only small engines to power it. Faced with a future full of possibilities, it has scope to become very successful.

But, like all European aircraft before it, the airbus must be accepted in America to have any chance of being a success in a commercial sense. Very early in its life, one of its first customers, Air France, took it to North America to use on the run from New York to the French islands in the Caribbean. With its range of around 2500 nautical miles, the airbus is not tailor-made for a crossing of the North Atlantic. The Air France aircraft, suitably stripped down inside, made it safely enough. But, mere presence in the States was not enough. It has required a concerted sales drive by Airbus Industrie to persuade the American airlines to take any interest at all. Western Airlines were the first to show enthusiasm but at the last minute it decided to buy an American alternative. Then Eastern, the airline whose chief is now astronaut Colonel Frank Borman, announced an interest in the European aircraft.

With its two engines and reduced fuel consumption, costing less per seat-mile to operate than either Tristars or DC10s, the airbus looked very attractive to Eastern Airlines. Borman, however, took an unusual preliminary step. He negotiated a rent-free trial of four airbuses for six months. Airbus Industrie agreed, and Eastern committed itself to some 7 million dollars for crew training and other costs associated with flying the new aircraft. Eastern has subsequently placed firm orders for more than 20 airbuses. That success gives the airbus not only a deal worth many hundreds of millions of dollars but also that vital foothold in North America.

So, the aircraft which is already in Europe, Africa, in the Middle and Far East, is now establishing itself in North America. The chances of you and me flying in it get greater all the time.

The success or otherwise of the airbus will have a fundamental effect on Europe's aircraft industry. The manufacturers themselves have no option but to plan on the basis that it will be a success and if there is to be a new generation of European airliners, it is likely to be based on a consortium like Airbus Industrie.

Plans for a smaller airbus are already well advanced with Britain no longer simply an outside sub-contractor but a full partner in the consortium. On 29 November 1978, representatives of British Aerospace signed an agreement with the German, French and Spanish members of Airbus Industries, taking a 20 per cent stake in the consortium and making an investment in the group which will amount to £250 million by 1983.

So Britain, whilst continuing to act as a sub-contractor to the American giants of the commercial aircraft industry, has taken what many people see as the only real chance left open to us to get back into the business of designing and building big passenger-carrying aeroplanes.

Auto Navigation

Earlier, I referred to the problems facing a German monorail. Let me now redress the international balance a little by sharing with you a remarkable experience I enjoyed in a factory compound outside Hildesheim in the same country.

I was seated at the wheel of a Mercedes enjoying the opulent feel of the big car. The area was strange to me, but at each junction I was given clear and concise instructions, not from a guide on my right, but from a small panel directly in front of me. With a slight bleep to attract my attention, the panel indicated with arrows whether I should turn or continue straight ahead, and also warned me of any impending fog, ice or traffic jams. When I arrived at my destination – and being a total stranger I had no means of recognising it – the panel informed me that I was there.

The system works using induction loops let into the road in the same way as diamond-shaped loops of wire are often laid in the road to detect vehicle build-ups at traffic lights. But these loops not only detect the passing of the vehicle

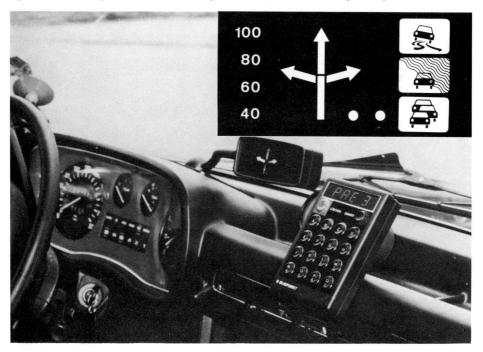

but they can transmit to that vehicle a direction signal for the next junction.

Inside the car, a small thumb-wheel counter is set with a number code for the destination. This is then transmitted digitally, from a small transmitter-receiver mounted alongside the front wheel. As the vehicle passes the induction loop in the road, the system picks up the destination code, processes it through a central computer, and transmits the instruction on how to reach that destination back to the vehicle. Each loop is around a metre and a half wide, so the whole transmit-and-receive sequence has to be completed in much less than a second, but, as I discovered, the system did not let me down.

Whilst attractive to the driver, the sytem is also of interest to the police and road authorities, who can see a build-up at some trouble spot. Everything will work successfully as long as the driver does as he is told. Remembering my firmly-held belief that I am the only one who can read maps, what happens, I wondered, if the driver takes no notice of the instructions?

Eventually, of course, he will get stuck in the traffic jam which the advised route was designed to avoid, but in the meantime the system will be telling him that he has gone wrong. If he makes an incorrect turn, the next loop will show him all three arrows on his display, pointing straight ahead, left and right. That is an indication that he is actually going away from, not towards, his coded destination point. This will continue until he turns round, and when he is heading in the correct direction once more, the system will pick him up and direct him accordingly.

Whether motorists in Britain would ever get used to taking direct instructions from a computer, particularly a computer programmed by a traffic policeman, must remain to be seen, but on technological grounds alone, the Hildesheim factory car park scheme is a genuine advance.

High Speed Rail

One means of transport we shall all be able to use in the coming years is the Advanced Passenger Train (APT). It has been a long time coming. The initial research programme started in 1964, but its appearance on our railways in 1980 will reflect British Rail's awareness that speed is good for business.

The story of APT really gets going in 1967, when the west coast main line from London (Euston) north to Liverpool and Manchester was electrified. The planners had predicted an increase in business as a result of the faster service, but everyone was taken by surprise at the way passengers flocked to the new trains. The airlines, which previously had been enjoying a very strong position amongst businessmen, had to launch an aggressive sales pitch to win back passengers lost to the fast electric trains.

British Rail's aim was to develop a very fast train. The west coast electrics are

The prototype Advanced Passenger Train – the fastest train to run in Britain

capable of top speeds of 100 mph. This new train would reach 150 mph, but even more important, it would do that over existing track. What little money was available for development would go into the train and not into a new track. Such a train was bound to take a long time to perfect, and remembering the lesson learnt in 1967 with the improved west coast electric schedules, a decision was taken in 1970 to build an intermediate train. This would not be as advanced as the 150 mph APT, but it would be an improvement on the conventional 100 mph. The hope was a slightly higher top speed that would bring even more passengers to the railway.

Just two years later, the prototype High Speed Train set up a new diesel world speed record of 143 mph an hour. For statistics freaks, that's 17 mph faster than the world record for steam locomotives set by Mallard before the Second World War. The first production trains came into service between London (Paddington) and Bristol and South Wales in October 1976.

The wide track bed and gentle sweeping corners, a legacy of Chief Engineer Brunel's decision to build the original Great Western line to a 7-feet gauge, suit HST well. Between London and Swindon, a distance of 73 miles, she can travel at top speed all the way, save for a reduction to 80 mph in the vicinity of Reading station. HST is not a particularly sophisticated beast. Her speed comes from her power, twin 2250-horsepower diesels producing nearly 3000 kilowatts of electricity to drive the traction motors mounted on the bogies.

She carries her power equipment divided between the front and rear cars of the train. This minimises damage to the track from one excessively heavy power car passing at high speeds. Between these twin power units are conventional coaches,

British Rail's High Speed Train

British Rail Mark IIIs. They are exactly the same as the ones which in less powerful trains are limited to 100 mph.

The other essential in any high-speed travel is the braking system. The electrically controlled air brakes on the HST actually enable it to stop more quickly from 125 mph than a conventional train travelling at 100 mph. And that is not just a reassuring safety feature; it means that HST can operate with exactly the same distance between signals as ordinary trains. No special signals are needed for HST.

The only other railway in the world to operate a high-speed service as intensive as the one now operating out of Paddington is the Japanese Tokaido route. To do that, the Japanese had to build a totally new railway exclusively for passengers who wish to travel at high speed and who pay specially increased fares. The philosophy behind both HST and APT is that they should cost no more to ride than any other train on the system.

Average speeds on the Paddington services now approach 100 mph, with HSTs regularly covering the 36 miles from London to Reading in 22 minutes – an average start-to-stop speed of 98.2 mph. That is the fastest schedule for diesel trains anywhere in the world. The HSTs are now also appearing on the east coast mainline out of Kings Cross. The stop gap between conventional trains and APT has become a success in its own right.

HST is also the end of a chapter, because it represents the limit facing engineers who want to make a train go faster simply by making it more powerful. Not that it would not be possible to mount bigger diesels and more powerful electric traction motors inside a conventional train's skin, but the higher speeds so attained would make life intolerable for the passengers. Half of Britain's railway track is curved and half of that is relatively sharply curved. A curve with a radius of 2 kilometres is sharp enough to justify slowing a train down. Not because the train will not stay on the track, but because the centrifugal forces working on the passengers, throwing them outwards like a stone on the end of a rotating string, would make life very uncomfortable for them.

Conventional trains can cope with this, because many curves are canted inwards to a maximum of around 4°. Where the APT differs from all trains that have gone before it in Britain is in its ability to lean inwards. This increases the tilt on corners to around 9°, raising the maximum speed through those corners by as much as 40 per cent. Without this trick of leaning into corners the APT's high top speed would be of little value because it would be forever slowing down for the next uncomfortable bend. What the railways now have is the promise of not only a high maximum speed, but also a high average speed.

But, such speeds bring problems. Aerodynamic drag increases dramatically in relation to speed. Drag needs power to overcome it, which means more engine, which means more weight, which means more drag, and so on.

An exceedingly streamlined low-drag profile has been used on APT. It is made up of smaller vehicles than conventional trains, with an aerodynamic nose and tail. Using this design, APT can travel at 125 mph, consuming the same energy as a conventional train would need to reach a hundred mph. Like HST, the APT can pull up from top speed, that is 150 mph, in the same distance as a conventional train from one hundred. This is achieved by a hydrokinetic system mounted inside each wheel axle. As the brakes are applied, the axles fill with a hydraulic fluid, and as the pressure on the fluid is increased, the axles find it increasingly difficult to turn round. The drive energy is transferred to the fluid which heats up, and is pumped to radiators for fan cooling.

The body shells of the passenger cars are aluminium alloy, some 40 per cent lighter than those of the conventional Mark III. Lightweight steel is used to house the power units. The first APTs which are scheduled to run between London (Euston) and Glasgow will be powered by 3000 kilowatts of traction motor. But the line from Euston to Glasgow will present problems to the new train. It is a particularly curvy route.

I remember, as a boy, wondering why the old London Midland and Scottish Railway Company would not match the London and North Eastern Railway at top speeds. The answer was that the former company, whose line to Glasgow APT will use, had nothing like the straight track of the LNER on which to run fast. On one occasion, when an over-enthusiastic LMS driver did try to break a record speed, he had to slow down so violently as he approached Crewe that all he managed to break was the crockery in the dining car.

Sadly perhaps, the LMS route will today demand all the cornering skill of APT to allow even 125 mph to be achieved. That is, do not forget, only as fast as the HST already goes between London and Swindon. But that is a considerable improvement on the present running speeds between London and Glasgow, and will reduce by a clear sixty minutes the time taken for this 400-mile journey.

Because it cannot corner as well as APT, an HST could only clip ten minutes off the current running time.

What will it be like riding on APT? The carriages will be small inside, because of the desire to reduce drag and because of the need to lean into corners. Britain's railways have a limited loading gauge – the maximum height and width of a rail vehicle able to use the system. It dictates particularly how tunnels and bridges are built, to ensure that there is clearance for passing trains.

Any carriage that leans into corners is going to have its width apparently increased as it goes round bends. It can only fit into a fixed loading gauge if the top of the carriage is trimmed down to start with. Consequently, the APT will have an unusual profile, getting narrower towards its roof.

There will be fewer doors in the APT, two in each carriage at diametrically opposed corners. The extra space will be used to house more seats to produce more revenue. And the early APTs will be the first British mainline trains in many years in which you cannot walk from end to end. This is because the engine, the power cars, will be in the middle. Passengers will be isolated, either in the front, or the rear halves of the train.

The engine in the middle is designed, amongst other things, to ensure that all the electric power can be collected from the overhead wires by one pantograph – the frame on top of the train. Even two power cars coupled together can collect from one pantograph, which will be easier to handle at high speeds than two pantographs at different parts of the train.

Even diesel APTs will adopt this same power unit configuration, because keeping the power in the centre reduces the buckling forces as the train picks up speed quickly. There is also less need for heavy coupling gear in the centre of the train, so more power-generating equipment can be included to improve the train's performance.

So it will be different from the trains we are used to now, and it will certainly be fast. An experimental APT travelled 5 miles at a steady 152 mph in August 1975. No British train had ever travelled so fast. The same train covered London (St Pancras) to Leicester in fifty-eight minutes. The fastest regular scheduled service at that time took eighty-four minutes.

But speed is only one consideration. Over-all efficiency is another. When APT begins its runs to Glasgow, it will be a low-power version capable of a top speed of just 125 mph, but also able to cope with gradients on Shap and Beattock of 1.5 per cent. It will carry only one power car with eleven trailers for passengers.

The eventual plan is to run a 150-mph APT with 2 power cars and 14 trailers. Such a configuration would mean extending station platforms – but it would give the advantages of high passenger capacity and high speed, and every passenger will be carried at the same cost-per-mile as the shorter low-powered version we are about to see.

A solar power station under construction in the United States

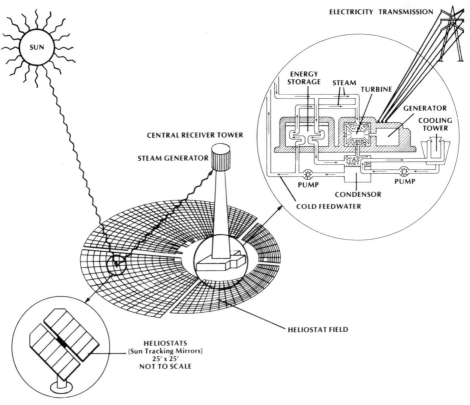

A solar power station. Sunlight is focused by mirrors to a central boiler at the top of a tower. Steam is produced which drives a conventional turbine generator housed directly below

at the top. One major problem still needs to be solved before the full potential of the idea can ever be realised. The sun moves across the sky; to extract the maximum energy, the mirrors will have to move as well. The mirrors will have to be steered continually throughout the day, to keep the rays aimed at the focus-point. This will need complex engineering, and it looks as though it's going to be very expensive. However, the American government is not deterred. It has funded four separate research projects with private industry, and the aim is to build the first solar-power generating station by 1980.

An alternative way of producing solar electricity is by using photocells. A cell is normally made of a thin slice of pure silicon crystal and a thin layer of metal such as silver. When light falls on the cell, electrons flow across the layers, and a small current is produced. This technology was originally developed for the space programme, but for earth use it's hopelessly uneconomic. At the moment cells to power an average-sized house would cost around £50,000, and of course they only work when the sun shines. But the search for the cheaper cell is on. In

America, research is centred around simpler ways of producing a silicon crystal. One method being tried is to pull a thin crystal from molten silicon. In Britain, one group is experimenting with glass. By using two layers, each sprayed with a mixture of chemicals, and then dipped in a copper solution, the group has managed to end up with a cell which will produce electricity. But making it in a laboratory is one thing. Being able to mass-produce it is still some years off.

Possibly the most way-out application for solar cells is to provide electricity for earth from power satellites. These would convert sunlight into microwave energy, which would be beamed down to earth and converted into mains electricity.

If the goal of sun-powered electricity is directly related to the amount of money invested in research, then the major problems could be solved by the 1980s; though whether the techniques will produce an economically worthwhile system is still anybody's guess. The major impetus for this work came from the 1973 oil crisis. Ironically, if the research succeeds many of the oil-producing countries will stand to gain the most. By the twenty-first century the Middle East could still be exporting energy – solar electricity.

Wind

Wind is one form of natural power that is actually used less now than it was 150 years ago, but the trend is about to change. In the past windmills were used to grind grain or to pump water for drainage and irrigation schemes. Their future role will be to provide electricity, and they are about to have a name change too; windmills are out, aerogenerators are in. But the principle of blades spinning in the wind remains the same.

Many of the new research groups looking into wind energy have an aviation background, so not surprisingly some of the designs have an aircraft feel to them. In Ohio, USA, engineers from NASA have built an aerogenerator capable of supplying enough electricity to power about twenty-five average-sized homes. It's an impressive sight. A tower 30 metres high, with an aerodynamically-shaped capsule on the top. At one end of the capsule are two long aluminium blades. The capsule is designed so that it always positions itself with the nose into the wind and blades at the tail. This machine starts to generate power in an 8-mph wind, but it doesn't produce smooth alternating current of mains frequency until the wind is up to 18 mph. At this stage the blades are spinning at 40 revs per minute. And that's their limit; even though the wind may get stronger, by varying the pitch of the blades, the 40 rpm is maintained, and with it the constant frequency electricity.

Similar machines are being tested in other parts of the world. But in the past, two-bladed structures have always suffered from problems of vibration; now, by using lightweight materials and wind tunnel tests, the designers hope to have produced a stable machine. But it is worth remembering that the world's largest aerogenerator, completed in 1945 on a hill at Grandpa's Knoll in Vermont, USA, failed after only one month's operation when one of its two blades broke off. Excess vibration had caused the metal to fatigue.

One answer is to use more blades. A private enthusiast in Yorkshire has produced a large three-bladed generator. He has linked the blades together so that they become more like the spokes of a wheel. With fixed-pitched blades he has produced a far less complex design than the large two-bladed systems, and yet his generator is producing enough electricity for himself with, on a windy day, a little left over for the grid.

Some engineers have gone back to first principles, and have come up with the original idea that all development so far has been going in the wrong direction. Instead of having the blades spinning round a horizontal axis, they claim it would be better to use a vertical axis. By turning the axis through 90 degrees the blades will accept wind from any direction, without the need to orientate into the wind. This is particularly useful when the wind is gusting, and its direction continually changing. A conventional machine will start hunting, trying to find the perfect alignment. With the vertical axis this problem is by-passed completely.

Several designs are being considered. The Darrieus, patented in 1931, is now being developed in the United States. It has a vertical shaft, with two thin semi-circular blades, each connected at the top and bottom. A three-bladed variant called the Troposkien is being tried in Canada. However, blades for these designs are proving difficult to manufacture. The problem is coping with the heavy stresses that build up when the blades are spinning very fast in a strong wind.

At Reading University, designers think they have a solution. They have produced a vertical axis machine with straight blades. It is 'H'-shaped, not unlike an old 405-line television aerial. The blades are hinged at their junction with the horizontal cross-bar. Above, they are supported by a tie wire connected to a spring in the shaft. As the speed of revolution builds up, centrifugal force causes the blades to incline, so that at very high speeds, the blades become an almost straight extension of the horizontal cross-bar. This ability to alter its shape enormously reduces the stress on the machine at high speed, while improving performance at low speed. It is claimed there is no wind too strong for this design.

While the search goes on for the ideal aerogenerator, meteorologists are collecting data to decide on the best sites. Isoplethic maps have been produced showing the wind energy available throughout the United Kingdom. Analyses have been made of the number of calm days, and the variation in the wind energy available between winter and summer.

Hill-top sites are favoured because they are always windier. The most likely reason is that the wind stream is compressed or focused as it is forced up the hillside, and the higher it goes, the more it accelerates. Britain's windiest hilltops

Vertical axis windmills under test in the United States

have already been identified, and a list has been drawn up of nearly 1500 suitable sites for aerogenerators. But will public opinion stand for 30-metres-high towers on some of the country's most picturesque hilltops? Also, each machine will need a line of pylons to carry the electricity to the National Grid, often a long way away. Another environmental black mark is that aerogenerators are going to make a noise; the tips of the blades will be moving very close to the speed of sound.

To get round these environmental problems, a proposal has been made to build 'wind farms' out at sea. The southern North Sea, off the Wash, seems suitable because there is a large area of very shallow water, much of it less than 20 metres deep. There shouldn't be any major construction problems as the technology of building structures at this depth has already been well developed by the oil companies. Just as important, there appears to be almost as much wind available as on those 1500 selected windy hilltops. It is proposed to build a cluster of 400 aerogenerators, each 500 metres apart, on a grid 10 kilometres square. On a very windy day the cluster would be capable of producing 1000 megawatts of electricity, but the output on an average day would be around 400 megawatts. And there is ample potential for expansion; there is enough shallow water to house up to thirty clusters.

A postscript: one researcher thinks there is a much simpler way of tapping wind energy. Alan Papesch of the University of Canterbury in New Zealand was studying the way trees behave in the wind. His research had been prompted by a particularly strong wind that blew over New Zealand's Canterbury Plains in August 1976. In the space of a few minutes this wind felled more trees than the lumberjacks planned to cut down during the next ten years. Papesch was looking into ways of ensuring that this devastation didn't happen again, and he built models of various forest layouts to test in wind tunnels. He was trying to find out why a row of trees would suddenly be completely flattened. However, the sight of his model trees gyrating in the wind tunnel made him realise he could kill two birds at once: stop the trees being blown down, and diminish the New Zealand energy gap. He was convinced that swaying trees represented a large amount of untapped energy. Since then, in a forest just outside Christchurch, he has rigged up a prototype pump, connected by a series of pulleys to four pine trees. As the trees sway the connecting wires either slacken or tighten, and this movement activates the pump. The next stage is to connect the pump to a small turbine. Alan Papesch calculates that electricity from swaying trees would be cheaper than from windmills because no structures would have to be built; also, with all the trees connected by a network of wires, there would be little chance of them ever being blown down. It seems too simple.

Hot Rocks

To some scientists, talk about a future energy shortage seems odd, when the earth itself contains all the heat that man could ever conceivably need. It is called geothermal energy, and it is in the rocks underneath us. How the rocks actually came to be hot is still the subject of scientific debate. Certainly some of the heat may be a hang-over from the formation of the earth, but it seems likely that most of it comes from the steady decay of radioactive elements like uranium and thorium deep in the ground.

The dual problems with geothermal energy are how to find it, and how to extract it. At the moment, the costs of drilling invariably make it more expensive to look for heat than to look for oil. Some places are lucky. The earth's crust can be very shallow, usually in volcanic regions. Heat can occasionally be seen escaping through cracks in the earth, sometimes as steam, or bubbling hot mud, or a geyser shooting up like a fountain. These sites are always a tourist attraction, but in Italy, California, and New Zealand, natural hot water and steam from the hot rocks are used to generate electricity. The Rift Valley in Kenya is another area where engineers are hoping to produce geothermal electricity. In Iceland most of the houses are heated geothermally. But these are all obvious sites. The existence of the hot springs and the like have meant that no searching was necessary. However, since the oil crisis, new geothermal sites have suddenly become worth looking for.

The New Zealand approach is typical. Geologists estimate that within their geothermal area there should be enough energy to provide half the nation's electricity. Every day, prospectors go out, and try to pin-point any spots where pockets of hot water are close to the surface. They consider any hot spot within one kilometre of the surface to be commercially worth exploiting.

For their water divining, they use a simple electrical technique. A 500-metre long wire is stretched out. At each end a steel rod is knocked into the earth. Then a strong direct current is sent along the wire, through the rod, and into the ground. The current tries to complete the circuit the only way possible, actually through the earth. Half-way between the rods, geologists check how hard the current has to work to get through the earth; they are measuring resistance. The lower the resistance the better the earth is conducting the electricity, and that means the greater the chance of water. And because the earth's crust is known to be thin, the water will almost certainly be hot.

A well is then drilled to test the steam pressure. An ideal site is where the pressure remains more or less constant. The hot rocks are continually heating rain water that has seeped into the aquifer. This hot water in turn escapes as superheated steam through the well head. The skill of the geothermal engineer is to drill just the right number of wells in any area. Obviously, over a period, the rocks must start to cool down but as long as the off-take is balanced by the rain-water seeping in, a well-managed geothermal site should last hundreds of years.

As the potential advantages of geothermal energy are realised, so more countries are trying to get in on the act. Germany, France, Egypt, Sweden and the UK are all mounting geothermal study projects. They are all drilling in areas where hot rocks are thought to be reasonably close to the surface.

The best sites in England seem to be in Devon and Cornwall, where the heat in the rocks is known to be due to radioactive decay. The temperature seems to increase by at least $35°C$ for each kilometre drilled. But the rocks are dry, so no natural steam will shoot out from these holes.

In the United States, pioneering research is being conducted on how to extract the heat from dry rocks. Geologists from the Los Alamos Scientific Laboratory in New Mexico have devised a method, which though still crude, could be the basis of a new energy industry. They drilled a hole into dry hot rock. At 3000 metres they had reached a temperature of $200°C$. They then injected water under very high pressure into the hole. This caused the rocks to fracture. The more fracturing they managed to create, the greater the surface area of hot rock exposed. The aim was to try to open up around 6 square kilometres of rock face deep underground. Then they drilled another hole about 75 metres away from the first, and repeated the rock fracturing process. Eventually they succeeded in creating some cracks which spread from one hole to the other. The next step was to fill the holes with water, which inevitably became heated by the hot rocks. They had succeeded in producing an artificial aquifer. More water was then pumped under pressure down the first hole. The only way it could escape was up the second hole. The water went into the ground at $30°C$; when it came back out, it was around $170°C$. Then it was passed through a heat exchanger before being pumped back down again.

Further trials are being conducted with this energy extraction system. More experience is needed in fracturing rocks and in learning how to connect holes 3000 metres underground. Skills will be needed in selecting the sites. If the bed of rock is too shallow the crack may extend to the edge when it is fractured. The next layer might be permeable and all the water would seep away.

The Los Alamos project is the world's first artificially created geothermal system. Research is still in its infancy, but by the turn of the century, energy from the rocks could well be a reality in large areas of the world.

One final geothermal thought. Some of the wells drilled in the North Sea have bottom-hole temperatures of around $150°C$. When the oil runs out all is not lost. The wells and the platforms could be ready-made geothermal installations.

Waves and Tidal Energy

If we could control the waves, we could throw every other energy project out of the window. Nowadays the energy experts generally agree that wave power is the one alternative energy project most likely to succeed.

In Britain we are short of sun, our winds blow intermittently, we have no natural steam, but we are well endowed with waves. The prevailing winds are westerly, from the Atlantic. By the time the waves reach our waters, they have had the benefit of being blown over a 3000 kilometre stretch of ocean. The potential power is prodigious. In a force 8 gale, a weather ship has measured that for short periods of time there is enough energy in each metre of waves to power 3000 one-bar electric fires. Even when figures are averaged over the year, the potential energy available, particularly off the west coast of Scotland, is still enormous. A barrage 250 kilometres long could provide at least a quarter of our present electricity needs. That is the challenge and also the problem. How to make an energy extraction device that will work and survive out at sea, even in the strongest gales.

A search of British patents shows that, over the years, there have been well in excess of a hundred proposals for extracting energy from waves. Flaps, floats, ramps and liquid pistons have all at some time been advocated. But like so many new energy projects, it needed the oil crisis to stimulate serious scientific investigations. British engineers, with full Government backing, now have a world lead in this research.

Stephen Salter, at the University of Edinburgh, is the pacemaker, and has been refining what he considered the ideal shape for a wave power extractor. He thinks it should float and have the profile of a pear, squashed flat on one side. His idea is to have a series of identical floats connected to a central spine. Each float will nod up and down as it is hit by a wave. Not surprisingly, he calls the floats 'ducks'. Each time the ducks nod, energy will be generated.

Much of the theoretical work is now completed. The shape has been tank-tested, and shows a remarkable efficiency. On calm days, the ducks can absorb 80 per cent of the available energy, though they are designed to be less efficient as the sea gets rougher. The problems still to be solved are primarily in the engineering, and in many ways these are the most difficult. What are the best materials to use? How will the ducks behave in rough open water? How should they be tethered? What is the easiest way of collecting the energy, and how is it to be sent ashore?

Hot on the heels of Salter is Sir Christopher Cockerell, the man who invented the hovercraft. He is now applying his talents to the problems of wave power, and has come up with a design that is basically a series of rafts, all hinged together. At each junction a ram-type pump will be fitted. As a wave passes under the unit, the relative movement of one raft to its neighbour will activate the pump. Electricity will then be generated by a fluid being pumped through a hydraulic motor, and

Engineers preparing wave energy 'ducks' for testing in Loch Ness

power will be extracted progressively as each wave passes under the rafts. It is thought that by using the relative motion of one raft to the other, no rigid mooring will be needed, and that should simplify construction.

There is no shortage of other ideas to investigate. The University of Lancaster is testing an air-filled sausage that is compressed as a wave passes along it. The National Engineering Laboratory is working on the pressure changes which can be created in an inverted, open-ended container as a wave passes beneath it.

When it comes to finance, wave power research has one major point in its favour. It is easy to start on a small scale, and as the problems are overcome, gradually increase the size of any device. Similarly, tests can start in sheltered water, before tackling the much rougher off-shore waters. The theory is that any fundamental problems should come to light long before the nation is required to pour vast sums of money into the project.

With tidal energy, the very opposite to wave energy applies. Each scheme is an act of faith. It has to be built full scale before even the most enthusiastic supporters can be sure it will work. The principle is simple. Build a dam across a river estuary, and use the difference in water level between high and low tide to drive a turbine to produce electricity.

The idea has already been tried out and proved to work. The French built a barrage across the Rance estuary near St Malo, in 1966, and have been generating a small amount of electricity ever since. The Russians commissioned a larger system 60 kilometres north of Murmansk in 1968. Both systems were planned as prototypes for something bigger, but in each case the decision has not yet been taken to go for larger schemes.

Like wave energy, we in Britain are well endowed with potential tidal power. The best site in Europe is the Severn Estuary, where the tidal range, i.e. the difference between high and low tide, is around 13 metres. The possibility of using this head of water for energy purposes has been appreciated for some time. In 1933 there was a proposal to build a barrage across the river slightly downstream from the Severn Bridge. The idea was never taken up. One reason may have been the threat to employment in the Welsh coalmines. In those days almost all electricity was generated from coal. Recently, new proposals have been made for a much larger dam. This would stretch roughly from Weston-Super-Mare to Cardiff, and would be capable of producing a tenth of our present electricity requirements.

The drawback with tidal schemes is that electricity is generated only when the tide is right, and that is not necessarily when people want the power. That problem has now been overcome. The latest proposals for the Severn consist of building two barrages – creating a secondary basin within the main one. If high tide is at a time when energy is needed, the sluices on the secondary basin will be open, and the system will work as if it is one large basin. However, if high tide is at a time when few people want electricity – for instance the middle of the night – the sluices are closed on the secondary basin. The primary basin works as normal, producing electricity. This power is then used to pump water up into the secondary basin, increasing its water level. It is like a giant battery – a way of storing energy. When the peak demand time comes round, the sluices are opened and the turbines between the secondary barrage and the open sea start to work regardless of the state of the tide. Neat, simple, but considerably more expensive than the single dam system.

The Severn Barrage is at least practical. Since the Dutch Delta scheme, when large sections of the mouth of the Scheldt estuary were dammed, the supporters have become convinced that there are few insuperable engineering problems. What is more questionable is the effect on the environment. Part of the Bristol Channel would be converted into a large lagoon, still to some extent tidal, but with a range of maybe only 5 metres. Much less water would flow through. Would this cause problems of silting in the dammed area? Would the upriver ports, particularly Bristol and Cardiff, be able to operate? And would the constant higher water level in the estuary have any detrimental effect on drainage from the surrounding land? Nobody can be sure that the problems have been solved until the system is fully working. Unlike wave power, tidal energy is more of a gamble.

Green Energy

Nature is an enormous converter of solar energy. The process of photo-synthesis, the means by which green plants use sunlight to grow, is now being re-appraised as a possible energy source.

Energy is stored by plants. Until recently, virtually every country depended on wood for its basic fuel. Even today, wood provides countries like Brazil and Kenya with over half their energy needs; Tanzania and Nepal are over 95 per cent dependent on it.

It is easy to see why. Trees grow abundantly in most places. Wood is simple to collect and no special skills are needed to burn it. In the modern industrial world, its disadvantage is that weight for weight it gives only half as much heat as coal, and only a third as much heat as fuel oil.

In America, though, planning is already well advanced on the forest of the future. This will be an energy farm, and will be run more like a farm than a forest. At present the biggest drawback with trees is the amount of time they take to grow. The energy farmers think they have got round this by growing their trees a different way. For a start, the trees will be planted much closer together. The forests will be so dense it will be impossible to walk through them. A conventional plantation may have up to 2000 trees per acre. The energy farmers have already tested 77,000 trees per acre, and at that density a plantation is distinctly more like a wheat field than a forest.

The forests will look quite different from today's vast conifer plantations. Conifers are softwoods, used mainly for pulp, and ideal for growing on marginal land. The new forests will have to be on good land, and the trees will be hardwood, like sycamore, poplar and willow. These have been singled out because when they are young they grow very fast.

Once the seedlings have been planted, weeds will be controlled and the seeds fertilised from the air. After five years the harvest starts. Using machines de-veloped from present-day combine harvesters, every tree on the plantation will be cut down, taking care not to damage the roots. They expect to harvest at least 25 tons of useful wood from each acre. The plantation will be fertilised again, and the trees will regenerate by coppicing, i.e. sprouting from each stump. And that is the big bonus of the hardwood tree over the conifer – the energy farmer doesn't have to replant for the next crop. After another five years the plantation will be ready to harvest again, this time with an even heavier crop. At the moment it is not known just how many crops can be taken, but the planners are thinking of at least a thirty-year life for the plantation.

What will the wood be used for? In Ireland they have already tried replacing peat with wood in a peat-burning power station with very good results. In the United States, where most of the research is centred, they are experimenting with a blend of wood and coal in power stations. But they also see the possibility of gasifying wood. Applying the latest gasification technology to wood, the engineers hope to produce synthetic natural gas. Another option is liquid fuel. Wood can be distilled to produce methyl alcohol (methanol) or wood spirit. Alternatively, the sugar which can be extracted from wood can be fermented to produce ethyl alcohol (ethanol) – a colourless inflammable liquid. Both methanol and ethanol can be blended with petrol.

To make ethanol as a petrol substitute is the objective behind the world's biggest energy farm project. In Brazil the Government has plans to try and switch all the nation's transport over to ethanol. At the moment nearly all their petrol is imported. By going for ethanol, they hope to grow all their transport fuel on Brazilian farms.

The plan is to refine sugar cane and cassava into alcohol. For years sugar-cane refiners have been producing small amounts of alcohol, mainly as a by-product in sugar manufacture. In future, special sugar-cane plantations will be set aside solely for fuel. Using present technology, 70 litres of alcohol can be produced from every ton of cane. Sugar cane gives very high yields, 60 tons to the acre, but it grows only in certain areas.

That is not the case with cassava – it will grow almost anywhere in Brazil. Cassava is a tuber and provides the basic carbohydrate in the Latin American's diet. It is familiar in this country in a granular form – tapioca. Brazil grows 24 million tons a year, a third of the world's production. And though the yield is much smaller than sugar cane, it produces two and a half times as much alcohol per ton. The Brazilians' main area of research is to increase the cassava crop yield. The target is to cover one per cent of Brazil with cassava, and together with the sugar cane, to produce enough ethanol to be completely independent of petrol.

In the north of Brazil is another untried energy source. The babacu tree grows wild, and produces a nut far harder than a brazil nut. Traditionally the nut has been split by hand, and the kernel sold for oil. Recently a machine has been developed that for the first time is able to split open the babacu, and this has resulted in a large increase in the number of babacu nuts being processed. Normally most of the protective layer around the kernel is thrown away, but it has been discovered that by heating this shell, it turns relatively easily into charcoal. Brazil, lacking coking coal, already makes steel using charcoal from wood. With millions of babacu trees, all producing nuts, the new source of charcoal is potentially enormous.

It is not only in Brazil that researchers think man has for too long ignored nut shells. In California there is a large walnut growing and processing industry. A pilot gasifier has been built to burn the waste walnut shells. The gas that is produced has a lower heat value than natural gas, but has proved ideal for small-scale heating in factories and farms. Using the same technology, trials are being

*Long bunches of nuts, suitable for charcoal,
hanging from a babacu tree in northern
Brazil*

planned to produce gas from corn cobs, rice straw and even tree bark. The
objective is to make small gas plants, serving a single farm or factory, using
products which are currently only considered as waste. Ironically, producing gas
from waste is a technology which was well developed in Europe during the
Second World War, but has long since been forgotten.

An even odder substance for making gas grows in the Pacific Ocean off the
Californian coast. It is a giant seaweed called kelp. Research into kelp gas is spon-
sored by the American Gas Association. Their need is clear. America has, over the
years, built up a large demand for natural gas and its supplies are beginning to dry
up. At the moment Americans are becoming dependent on imported gas, and the
need to find another indigenous source is paramount.

Californian giant kelp is one of the fastest growing plants in the world. Under
ideal conditions it can grow over two feet a day. The plan is to have kelp farms
anchored about 10 kilometres offshore, in about 160 metres of water. A network of
underwater floats will be set up, 30 metres below the surface, rather like an
inverted umbrella frame. A single kelp plant will be transplanted on to each float.
Water near the ocean bed is known to be much richer in nutrients than water
nearer the surface. The intention is to pump up this deep water directly into the
kelp beds. The kelp will naturally grow upwards towards the daylight, and a
regular supply of nutrient-rich water should keep it growing flat out. On the
surface, harvest boats will mow the kelp. Once the top is cut off, the kelp acts like
grass; the sunlight regenerates the plant.

The kelp will be taken ashore and dried, then fed into a digestor, and allowed to ferment. Specific micro-organisms will be added to break down the kelp, producing the all-important methane. This gas will be so pure that it will need little further processing before being fed into the gas distribution network. The remaining sludge in the digester will be used as a fertiliser, either commercially, or recycled by the harvest boats back to the kelp farms.

Making gas from seaweed may seem a far-off possibility but much of the initial groundwork has already been done. Natural kelp beds along the coast of California are mown regularly by harvest boats, and the kelp is used by the chemical industry. An experimental farm has been set up off the coast, and kelp successfully transplanted in it, though it was subsequently destroyed in a storm. The technology of fermenting kelp to produce gas is known to work at least on a small scale though the ideal micro-organism has yet to be selected.

The big unknown is the engineering. Can a structure be built at sea to support the kelp on the scale needed to be economic, and be strong enough to ride out bad weather. Also, can a simple system be developed to circulate (they call it upwelling) the nutrient-rich water from the ocean floor up into the kelp beds?

The latest design for a seaweed farm. The frame holds the Californian kelp plants. Nutrients come from deep-sea water pumped up through the central spine. Boats will mow the weed as it reaches the surface

Infra-red photo of central Oxford. The lighter buildings are losing the most heat. The warmer River Cherwell shows clearly at the top right

The opportunities for natural power are all around us, but it's unlikely that many will be developed commercially before the turn of the century. And yet in the long term they have to succeed, because natural power, unlike oil, coal, gas and uranium, is renewable. It will never run out.

Conservation

No chapter on energy, albeit renewable energy, is complete without some reference to conservation. In Britain, like all industrial countries, the scope for cutting down on the wasteful use of energy is vast. Are you saving it, or wasting it? The development of infra-red photography means that it is now possible to spot any house that is throwing energy away.

Radiant heat produces invisible electromagnetic radiation with wavelengths between visible light and radio waves. Infra-red photography takes an image of this radiation. The picture produced depends on the heat of an object. It means the outside temperature of a building can be examined instantly. A roof that is well insulated will look a different colour from one that has poor insulation. Double-glazed windows will show up, as will draughty doors. And if a house has central heating it will be possible to see if the temperature is kept too high.

Mount the camera in a helicopter and it will be possible to fly over an estate and pick out those houses that are wasting heat through the roof. But its use in industry will be more valuable. In a large factory complex it is often not possible to work out where heat is wasted. Open doors, broken windows and badly insulated pipes will all show up on the photo.

An aerial survey, looking for heat going to waste, is planned for main centres in Britain, and the first target is going to be industry. But cutting out waste doesn't always mean shutting doors and windows, and keeping the rooms at a lower temperature. It can mean putting waste heat to use. Usually, all that is needed is lateral thinking – seeing an opportunity to do more than the obvious.

Take the case of Liverpool's main newspaper, *The Liverpool Daily Post*. A new building was wanted; one that could house the printing presses, the newspaper offices, and have spare office space available to let. The architects designed a thirteen-storey office block built on top of the printing works. The design had one big selling point; it was to be air conditioned to keep it pleasantly warm all the year round, but more than likely it would result in no heating bills whatsoever.

The designers used their ingenuity and it paid off. A modern printing press has thousands of moving parts. When they are working, because of friction, they all produce heat. In Liverpool, the machines are being used most of the time, printing morning and evening papers. Normally an elaborate ventilation system would be installed, but even so most of the time the doors and windows would be open to cool the works down. That is not the case in the new building. Here the warm air from the presses is sucked into heat exchangers and that heat is then available for the office block above. There is more than enough heat to keep all thirteen floors warm.

An even more unusual example of getting something for almost nothing is at the Glengarioch Whisky Distillery in Oldmeldrum near Aberdeen. The distillery produces around half a million gallons of top-quality malt whisky every year. In any whisky distillation process, vapour is given off, and as this cools down, it condenses into whisky. To speed up the condensation, cold water is circulated around pipes containing the vapour. Over 12,000 gallons of water flow out of the distillery every hour, warmed up by the whisky-making process to about 65°C. In the olden days this used to be poured straight into the local river, but nowadays Glengarioch, like most distilleries, pumps the warm water through a cooling system and then recycles it back into the condensers.

It was when they wanted to expand production, and needed to be able to cool more water, that they had an idea. Why not make use of the warm water? They built a greenhouse, covering half an acre, alongside the distillery. They planted tomatoes inside. By piping the warm water from the distillery around the greenhouse, they have entered the market garden business without any heating costs. It works well, but tomatoes are only the start. They see no reason why they shouldn't grow melons or even pineapples.

All over the country, industry has installed equipment to cool down process water. And at the same time market gardeners are having to face enormous heating costs. After the pioneering work in Scotland, we may find that Birmingham and Manchester become centres for horticulture as well as industry. In the future a greenhouse could provide every factory with its own green belt.

But industry's problem of cooling down water is trivial compared with the Electricity Generating Board's problems. Every major power station has to have either enormous cooling towers, or be sited by the sea, merely to get rid of warm water. Millions of gallons have to be cooled every day. Even at the most efficient power stations, more energy ends up warming waste water than in generating electricity. Almost every large power station built during the last ten years wastes enough heat to provide every house in a city the size of Birmingham with constant hot water and full central heating. And yet the planners claim this is the sensible way to make electricity. But they don't think like that everywhere. In Sweden and Germany many of their power stations produce slightly less electricity, but their waste water is much hotter. Hot enough to sell. The result is that in many towns there is a network of hot-water pipes radiating out from the power stations. Every new house is encouraged to connect into what is called the district heating system. In Malmö in Sweden, around three-quarters of all the households are connected to the network.

Suggestions for similar systems in Britain have always been turned down as being uneconomical. Could it be that our energy is still too cheap? Conservation will never be taken seriously until energy costs are so high that people really begin to think of the cash value of the heat going up the chimney, out of the window, or into the cooling tower. Even though costs have increased, they are not yet high enough to encourage large-scale conservation. Ironically, if natural power lives up to its promise, conservation may never really need to catch on at all.

William Woollard

Inner Space

A little over a decade ago, at a typically stuffy bureaucratic meeting of the United Nations Assembly in New York, a delegate who until then had had little impact upon the world's affairs stood up to make a speech that has assured him a niche at least in the maritime history books. The man was Arvid Pardo, Ambassador for Malta. The speech was about the oceans of the world. Specifically, Pardo proposed that 'the seas beyond the limits of national jurisdiction should become the common heritage of mankind'.

Three years later in 1970 the idea was officially embodied in a set of Principles drawn up by the General Assembly. Some observers called it the high point of United Nations history. Since then, unfortunately, it has been mostly downhill.

For some nations – particularly those of the Third World and those without access to the sea – this speech was a ringing clarion call that heralded an end to many of their crushing problems. Many of them reasoned that if they were joint owners of this vast area then somehow – in a way as yet unexplained – science and technology would eventually gather from it a harvest of resources rich enough to meet all their needs.

For others, particularly the industrialised nations of the world, who either had the technology for getting at some of the wealth beneath the oceans or at least the means of developing that technology, Arvid Pardo's phrase was fraught with problems. Nobody had paused long enough to define what it meant. Did it mean 'common ownership' or just 'a common right to exploit'? Did it mean that those who had the technology now had to wait around until all nations had the technology to mine the sea-bed? Or did it mean they could go ahead with their plans for exploitation as long as they paid a fair slice of their profits into some nameless international trustee body? What's a fair slice anyway?

They have been arguing about it ever since. Indeed the argument goes on at yet another session of the United Nations Law of the Sea Conference (UNCLOS) even as I'm writing this. Perhaps it was all too idealistic anyway, bearing in mind the unequalled scale of the problem and the vastness of the prize. It represents the most stupendous share-out that man has ever seen; the allocation of no less than two thirds of the earth's surface, roughly equivalent to sharing out the entire surface of the moon between the 158 nations of the world.

As one international oceanographer has put it:

'The prosperity and happiness of the entire world may depend on what happens under the sea in the next decade. From the depths of the sea we may

This is how the ocean floors have evolved in the 200 million years since the break-up of the mighty continent, Pangaea. The mid-oceanic ridge shows up clearly. The white squares show some of the Deep Sea Drilling Project sites

harvest enough food to make the difference between getting by and enduring ·mass starvation. We may extract enough oil and minerals to keep industrial societies going and permit industrial benefits to come to the developing countries, or we may quarrel so bitterly over these resources as to trigger off another world war.'

So far most of the manoeuvring for position has taken place verbally around the UNCLOS' negotiating table, and as if the issues were not complex enough they have been further complicated by the grouping and regrouping of nations around three basic axes. There is the traditional East–West confrontation, between Russian dominated and American dominated groups. Then there is the so-called North–South split between the industrialised 'haves' and the non-industrialised 'have nots'. The third confrontation slices across the other two: between the sea-rich nations and the 'land-locked and geographically disadvantaged nations' as they are called in the jargon of the conference. This split puts some strange fellows to bed together; America, Russia, China, Chile, and India on one side, for example, and East Germany, Austria, Chad, Bolivia, and Zaire on the other.

But while the diplomats and the international lawyers were immersed in the minutiae of negotiations, the research ships were at work, sounding, measuring, sampling, testing prototype dredges. Just five nations have the technology to begin to scratch at the wealth on the sea-bed: America, France, Germany, Japan, and the UK. All are inextricably bound together in five major international consortia which are racing one another for a first-generation mining system. One of the companies has actually staked out its claim to 60,000 square kilometres of the bed of the Pacific Ocean, just to the north-west of Hawaii. Since there was no officially recognised procedure for claiming exclusive mining rights over an area that is literally in the middle of, and 5000 metres under, the surface of the Pacific, they did so by filing their claim with the State Department of America, with the United Nations, with every head of state of a recognised nation, and with all the leading mining companies. The only record of any response is that from the American State Department acknowledging receipt, although I'm told that the British Government replied with a somewhat curt rejection of the claim.

But bureaucratic scorn is a somewhat frail barrier to put in the path of a gold rush. Arvid Pardo raised the possibility that the last great resource on earth, the ocean's riches, could be used like a vast, unexpected legacy, to go some way towards redressing the imbalance between the rich nations and the poor. But as with every other gold rush in man's history, it looks very much as if it's the greedy, the ambitious, and above all, the first on the scene, who are going to make a killing. The rest may well be left eating sand.

The Ocean Bed

The ocean bed is strangely symmetrical. For all its mystery and uncharted depths, if you were to drain the oceans of the world and travel across them from coast to coast, you would traverse virtually the same features in very much the same sequence.

Next to the shoreline there is the continental shelf – broad, flat, approximately 200 metres in depth, but ranging in width from a few kilometres, as off the coast of Chile, to 300 kilometres or even more, as off the coast of the Eastern United States.

Then the sea-bed drops away into the continental slope which drops down to as much as 3000 or 3500 metres, often scarred and cut by deep canyons and valleys before it rounds out into a broad, gently sloping alluvial area known as the continental rise. This marks the outer limit of the continental margin. The broad flat abyssal plain is 6000 metres down, and stretches away for hundreds and thousands of kilometres. Then steadily the ocean floor begins to rise until it becomes a huge mountain chain – deeply ridged and faulted, several hundred kilometres wide and over 60,000 kilometres long. Known as the mid-oceanic ridge it is a vast interlinked chain of mountains that circles the world in the centre of all the major oceans. At its peak it is split by a deep cleft, which plunges down into the molten magma beneath the earth's crust and through which there is a constant upwelling of new molten material. This is the engine room of the earth's tectonic plates; the dozen or so giant pieces of jigsaw that make up the earth's crust and are ploughing into one another at the rate of 2 to 10 centimetres a year.

On the other side of this mid-oceanic ridge, the mountains slowly fall away again to the abyssal plain and the pattern is repeated – abyssal plain, continental rise, slope and continental shelf – before you find yourself back on the beach on the other side. The journey across the Pacific or the Indian Ocean might be longer than that across the North or the South Atlantic, but the view would be very much the same.

Who Owns What?

Ownership of the sea-bed is a modern invention – one of those revolutions that has occurred almost without our noticing and directly as a result of pressure on the resources of space-ship earth. Until as recently as the early sixties the only boundary line drawn on the map of the oceans was one that owed its origins to custom and practice – the practice of firing cannonballs to defend one's shores against seaborne attack. A cannonball could, it was reckoned, fly about three miles. So that narrow band just off the beach came to be regarded as territorial waters, owned and policed by the countries fortunate enough to have a coastline. In the rest of the deep, the principle of 'freedom of the seas' has been defended fiercely for many hundreds of years – particularly by those maritime Super Powers who had most to gain from it. That is, until 1958, when suddenly the ever-extending predations of the Russian and Japanese fishing fleets raised the possibility of fishing stocks becoming depleted. Even more critically, oil reserves were seen to be less than infinite and countries began to lay claim to the as yet undiscovered reserves under the sea.

The old order rapidly crumbled – to be replaced by an equally inefficient new one. In 1958 the Geneva Convention gave birth to what seems an extraordinarily short-sighted piece of international legislation. It gave coastal states exclusive ownership of the natural resources locked up in their continental shelves which it defined as '. . . the submarine area adjacent to the coast . . . to a depth of 200 metres . . .' So far so good, but then it goes on, 'or beyond that limit to where the depth of the overlying waters admits the exploration of the natural resources of the said area.' Within twelve years the research ship, *Glomar Explorer*, was drilling in 6000 metres – and by this definition the coastal nations could, perfectly legally, extend their ownership of the sea-bed until they met in the middle of the oceans. Indeed a map has actually been prepared to show what this would mean. It is called the World Lake Concept and it would give America a little under half the entire planet.

That briefly is the law as it stands now. Indeed, this point was recently invoked in a very blunt American speech in an attempt to break the grip imposed on negotiations by the Group of 77. 'Remember,' America's spokesman warned, 'we have under existing international law, including the 1958 Convention on The Law of the Sea, the perfect right to extract mineral resources from the deep ocean.'

What's Down There?

Just over a century ago the British research ship *Challenger*, on a four-year cruise around the world, recovered some small black potato-shaped stones from the bottom of the Indian and Pacific Oceans. At the time, and indeed for the next three-quarters of a century, the discovery did not attract much notice. Even today, no one really knows how these nodules are formed. It is thought that somehow salts and minerals are slowly deposited from the surrounding sea water, collecting around some nucleus – anything from a shark's tooth to a volcanic pebble.

By the late fifties it was realised that vast quantities of these nodules are spread over millions of square kilometres of the sea-bed. They lie exposed, or very lightly covered by the sediments. The richest areas appear to be in the North Pacific, but there are extensive deposits off South America and Australia, in the Western Atlantic and the Indian Ocean.

The nodules can contain a very wide range of minerals: iron, manganese, aluminium, nickel, copper, lead, zinc, titanium, molybdenum, cobalt, and more besides. But the minerals that excite the greatest commercial interest at the moment are the nickel, copper, and cobalt, with possibly manganese and titanium running a close second. The reason? Well, these minerals are critical to the economies of all the major industrialised countries, they're expensive and, most important of all, America, Germany, Japan, France, and the UK rely almost entirely on imports. The nodules could make them independent of threat, sanction, or price rise. America, for example, expects to be paying $7000 million in import bills for four of these minerals alone.

How long would the nodules last? Well, estimates vary from several years to forever. Although the rate of growth is something of the order of a centimetre per thousand years, several people have calculated that this, multiplied by many thousands of millions of individual nodules, actually represents a rate of accumulation that is greater than the likely rate of consumption. So, on that sort of reckoning, the nodules represent an inexhaustible source of wealth.

But every dream has its rude awakening and the nodule miners seem to have reached theirs particularly early. Just as the world's major oil companies never miss an opportunity for telling just how costly and risky a business looking for oil is, so all the major deep-sea mining consortia (several of which have oil companies as members) are already making it known what a marginal business it will be.

What has certainly become clear is that, unlike the dreams of the late sixties, we're not going to see fleets of mining vessels cruising gently up and down the sunlit surface of the Pacific scooping endless quantities of mineral wealth into their holds. It is going to be a long, slow, difficult, horrendously expensive operation. Around $200 million has been spent over the past four or five years, in the prospecting phase, looking for 'mines'.

A viable mine has been defined as an area of the sea-bed, perhaps 20,000 square kilometres, that would sustain a dredging operation for at least twenty years. It would have to yield each year around three million dry tonnes, not of any nodules, but of those with a combined nickel, copper, and cobalt content of 2.25 per cent or more. And even in this prospecting stage, the industry has had to develop its own technology. Sleds are towed 6000 metres down across the abyssal plain, carrying lights, still and television cameras, and highly accurate acoustic sounders to maintain the sled's position just a few feet above the mud. The French in particular have made great use of a new, free-fall sampler. The survey ship makes a run on a predetermined line, dropping the samplers over the side at set intervals. The samplers sink steadily towards the ocean-bed with their wide aluminium jaws held open. As they hit the sea-bed the impact slams the jaws shut, ballast weights fall off, and the now positively buoyant samplers return to the surface, hopefully with a bunch of nodules trapped in their jaws. The survey ship steams back along its track, collecting the samplers one by one.

The results of all this intensive survey work have been mixed. There aren't, it seems, as many nodules as the ocean mining mania of the sixties would have had us believe.

The nodules also vary enormously in quality and content even over relatively short distances. But they are undoubtedly there, in their manifold millions, and the potential rewards for a really efficient mining system are colossal. At the moment there are two main recovery methods that have been developed and that are being tested in ever deeper waters.

The 'continuous line bucket' system (CLB) is a French development of a Japanese technique. Basically it requires two ships, steaming parallel, perhaps a kilometre or so apart. Ship A lets out a line with buckets fixed at roughly 30-metre intervals. The buckets stream down to the sea-bed and scoop up the nodules, ship B winches the line in, empties the nodules into its capacious hold, and returns the line of empty buckets to ship A. Very simple, very neat, and relatively in-expensive as space-age technologies go – in theory. In practice the CLB has run into immense difficulties. In several of the test runs they have spent most of their time untangling the immense spaghetti pile of cable. Then there is the problem of attaching the buckets to the line in such a way that when they are 6000 metres down they fall at just the right angle to scoop up something off the sea-bed, even if it is only sand. You can't afford to have more than the odd bucket returning to the surface empty after a 12-kilometre journey there and back.

For a long time now the other consortia have piled scorn on the CLB system, describing it quite openly as a non-starter. It is true that during trials in 1977 the CLB system ran into trouble and was able to scoop no more than a few tonnes of

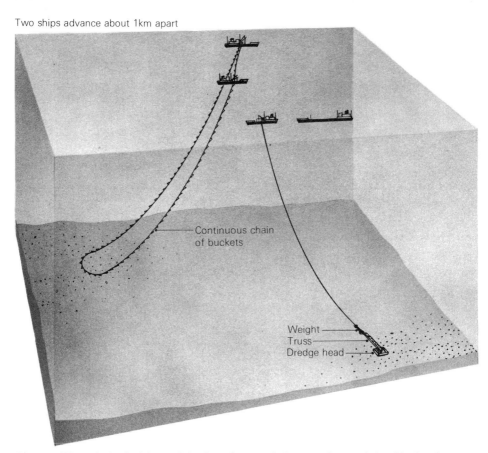

Continuous chain
of buckets

Weight
Truss
Dredge head

Two possible methods of mining nodules from the ocean bed – a continuous chain of buckets between two ships and a suction dredge

nodules to the surface. However, later tests in 1978 have been far more successful and the French will be taking their system out to the deep ocean by midsummer. It could even be that they will come from behind and literally scoop the pool because the other systems continue to show signs of trouble. These are the 'suction dredge' systems.

The mining vessel tows an immense steel pipe 5–7000 metres long, curving away beneath it through the darkness of the sea to the ocean floor. At the bottom of the pipe there is a sled 3 metres wide, designed to run across the sea-bed sediments, carrying lights, television cameras and a suction head. It functions rather like a giant vacuum cleaner. The nodules are scraped up, sucked into the pipe and forced up to the surface either by a sea-bed hydraulic pump or by air lift; the latter works by injecting air at intervals along the length of the pipe into the mix of sea water, sediments and nodules in order to reduce the density of the mix sufficiently for it to keep on rising.

This system has already been tested successfully down to 1000 metres. It is being tested now in conditions much closer to an actual mid-oceanic mine in the

Pacific and although for a long while it looked like being the first system that would go into operation, deep-sea tests in the spring of 1978 by two of the leading mining consortia had to be cut short. It is clear that there are still fearsome problems yet to be solved. It is worth remembering that up till now, the most successful, deep suction dredge in the world was pulling sand and gravel up through just 45 metres of water. The weight of the 6000 metres of pipe alone puts the supporting mechanism right at the limits of steel technology. If they get the air lift system even slightly wrong they could, as one engineer put it, 'be sucking manganese nodules off the sea-bed only to accelerate them into near earth orbit'.

But technological problems apart, just pause for a moment to consider the operational difficulties. The mining vessel will use satellite navigation to position itself to an accuracy of plus or minus 50 metres on the surface of the Pacific, but then it has got to track backwards and forwards across a given area with something of the accuracy of a tractor ploughing a field, only in this case the ploughshare will be 6 kilometres away, and the 'tractor' will be bouncing about in the wind and the waves. At the same time the suction dredger will have to transfer 10,000 tonnes of nodules a day to a nodule freighter towed along behind. Not an easy proposition.

Those are some of the reasons why they are hoping to achieve something like 20 per cent sweep efficiency from these early systems; that is, recover 20 per cent of the nodules in a given mining area from the sea-bed.

It is for this reason that l'Association Française pour l'Étude et la Recherche des Nodules (AFERNOD) has already gone ahead to design a far more adventurous second-stage system for which it claims a 78 per cent efficiency. Basically, it consists of a huge floating 'mother' platform rather like a gigantic semi-submersible drilling rig, equipped with a fleet of fourteen medium-sized un-manned sea-bed shuttles. These craft, completely automatic and powered by a clutch of lead acid batteries, will dive off to the sea-bed and crawl around on twin Archimedean screws dredging up the nodules.

They will navigate automatically in response to a sea-bed transponder net-work, and then return to the platform with their hoard of over 100 tonnes of nodules. Each trip will last six hours, each shuttle will do four trips a day. The platform will also be equipped with a processing plant, so the waste will be returned to the sea-bed in the shuttle, while the partially refined ore will be loaded into carriers and shipped to the mainland processing plant. But that word 'processing' should not be allowed to slip by unnoticed.

Not only is it of course a critical stage in any full-scale mining operation, but it has also proved to be one of the knottiest problems to crack. Basically there are only two ways to get at the minerals locked up in the nodules: to melt them, or to dissolve them in acid and extract the minerals chemically.

One consortium has tested at the research level over 100 different combinations of these two basic methods. The problem seems to be that the minute traces of thirty or more metals contained in the nodules interfere with any simple straightforward extraction process. But whatever the problems there is no doubt that within a decade, the biggest and most expensive mining operation the world has ever seen will be under way, and all the major industrial nations of the world will, as a matter of course, be getting some of their vital minerals from the deep-sea-bed.

What else is down there?

What else is there down there? The literal answer to that might be 'everything, absolutely everything'. The sea is a complex mixture of dissolved salts. Gold, silver, tin, zinc, copper, uranium, and several hundred other valuable elements. It works out at between 30 and 40 million tonnes of metal salts for every cubic kilometre of sea water and someone has even attempted to put a value on that lot. It comes out at $6000 million.

Is it ever going to see the light of someone's bank account? Well, only a few months ago a highly respected British University Professor of Geology was talking of special ionic transfer filters that would be towed behind ships, sucking the gold and other metals out of the waves. The same man also talked of the rich, solid rock, ore bodies yet to be discovered on the continental margins. Not too many of his fellows agree with him. But there are some fascinating cachés of treasure that are known to be down there although so far they have proved somewhat in-accessible. Diamonds are strewn at random across the sea-bed off the south-west coast of Africa. A sampling device, still at work, has been picking them up in ones and twos for years, but so far they have eluded any operation on a commercial scale.

Gold is there too, in thick veins in the sands of the continental shelf, but no one has yet devised an economic submarine prospector. Then there are the mysterious hot brines and sediments that lie in troughs deep below the surface of the Red Sea, fabulously rich in zinc, copper, silver and gold. A deep-water El Dorado if ever there was one, worth billions of dollars, and one that will certainly be exploited in the foreseeable future.

Yet, however you look at it, at the moment there are just two ocean mining industries that outweigh all others in scale and wealth. One is sand and gravel; it may be mundane but it is bigger in England than the National Coal Board, it has been profitable for decades and in terms of tonnage it is by far the greatest mineral harvest man is ever likely to take from the sea. The other of course is oil.

During the 'exploration phase' the research vessel Prospector *brings aboard another haul of manganese nodules from the bed of the Pacific*

The Glomar Challenger, *mother ship of the Deep Sea Drilling Project*

The Widening Search

In the middle of February 1970 the *Glomar Challenger*, a deep-sea drillship, was poised over a group of dome-like structures rising up out of the smooth floor of the Gulf of Mexico. They are known as the Sigsbee knolls. Four thousand metres down, the drilling bit ground its way through the cap rock overlying one of these knolls. Very soon it became clear from the examination of the drilling cores that they had located a reservoir of oil, trapped between a salt dome structure, and the cap rock. The drilling bit was withdrawn, the hole was plugged with cement and the *Glomar Challenger* raised its anchor and moved on. The scientists on board were excited over the discovery of salt at such great depth and the controversy that would arise in the discussions over the origins of the Gulf of Mexico. But for the major international oil companies it was the whiff of oil that rocked them back on their heels; oil from the deep-sea-bed. In many ways however, the Sigsbee knolls are something of a freak.

The big oilfields of the future that every major nation and oil company is searching for with unprecedented intensity are not going to be found in the deep oceans. The rocks are relatively young and mostly composed of porous volcanic basalt which has spread from the mid-oceanic ridges. Not the stuff under which oil is trapped. If they are to be found at all it is likely to be on the still unexplored areas of the continental shelf, such as the vast plateau that extends from the Siberian mainland for over 1000 miles northwards; and on the continental slope, down to about 3000 metres. Already the oil companies are girding themselves for a task that will exceed even that of exploring and exploiting the notoriously tough conditions of the North Sea.

At the moment there are only two kinds of drilling structure that can get off the edge of the continental shelf into deeper water. One is the basic semi-submersible: a platform mounted on wide cylindrical legs 100 metres or more high, that are themselves mounted on large, hollow flotation chambers or pontoons. Despite their vast size most semi-submersibles have some sort of power unit so that they can move under their own steam although they would normally be moved from site to site with the assistance of a flotilla of tugs. While it is moving, the pontoons act as the stabilising keel, but once in position the ballast tanks are flooded and the whole platform settles deeper into the water, leaving about 15 to 20 metres of the main legs above the water line. The buoyancy comes mainly from the pontoons that are now deep underwater below the main influence of the waves, so that even in heavy seas a lot of the pitching and vertical movement is damped out. But even though the semi-submersible is heavily anchored all round to keep it from drifting and so damaging the drill string, in really heavy weather drilling operations have to stop and the drill is pulled back on board. There are some eighty semi-submersibles working in various parts of the world, but even these monsters have their limitations. None of them has yet operated in much over 300 metres of water. On the continental slope the requirement will be ten times deeper. The answer would seem to be the drillship, and here undoubtedly the *Glomar Challenger* has blazed a most remarkable trail.

This small, even slender ship (she displaces only 10,000 tons) is 400 feet long and 65 feet in the beam, and is totally dominated by the giant drilling mast mounted amidships that soars 194 feet above the waterline. As the drillship of the Deep Sea Drilling Project, run by an American-dominated but now international group of research institutes, she has drilled almost 1000 holes in the floor of every ocean in the world. Over the past ten years she has provided scientists with a vast avalanche of data on the nature and age and history of the rocks that make up the sea-bed, and on the way the earth has evolved over the past 150 million years. It has been described as the most successful scientific experiment ever carried out, but its range of success goes well beyond the world of pure science. The *Glomar Challenger* has pioneered drilling in the open seas in waters as deep as 6000 metres, and for this she has achieved some major technological breakthroughs.

The ship had to be able to maintain its position precisely over the drilling site despite wind and waves and sea currents. Anchoring in several thousand metres of water was out of the question, so they pioneered dynamic positioning. In addition to its main propellers, the drillship has twin pairs of thrusters, smaller propellers, housed in narrow tunnels mounted on both sides of the hull, fore and aft. Underneath the hull, again both fore and aft, are sensitive hydrophones. The hydrophones, the main propellers and the thrusters are tied in to the ship-board computer. When the drillship reaches a drilling site it first positions a sonar beacon on the sea-bed and then switches the computer into the dynamic position-ing mode. The computer uses the constant stream of signals from the hydrophones to determine its position in relation to the sea-bed beacon, and then drives the thrusters and the main propellers to counteract the effect of wind and waves. In 6000 metres of water the system will keep the ship inside a circle 30 metres across.

Then there was the re-entry problem. When the Deep Sea Drilling Project started out only one drilling bit could be used in drilling the hole. As it cut through the rock layers underlying the sea-bed sediments, its cutting teeth were steadily worn away and when they had gone, that was the end of that bore hole. They had to move on to the next site.

The drilling 'string', that is the drilling pipe that runs from the derrick on board ship, down through the 'moon pool' amidships, and on down to the sea-bed, is often thousands of metres long. Once they had pulled that drilling string out of the bore hole, they could not change the bit and then get back into the hole. It was proving to be a very severe research limitation so the engineers attached to the project got to work to develop a remarkably successful re-entry system. When it first starts to drill a hole in the sea floor the drillship locates in the hole a giant cone-shaped funnel, around the upper lip of which there is a ring of sonar reflectors. When the first drilling bit is worn out, the drilling string is pulled back on board, the drilling bit is replaced and then they attach a mini sonar scanner to it. As the string is slowly lowered away through the moon pool, the sonar scanner sends out a series of pulses and picks up the echoes that come back from the sonar reflectors positioned around the funnel on the sea-bed. On board ship, the operator has two cathode ray screens. One tells him where the ship is in relation to the hole, the second where the drilling bit is in relation to the funnel. By careful adjustment of the rate of descent in relation to the swing of the drill pipe, he can guide the drilling bit back into the funnel.

Another problem encountered by the *Glomar Challenger* was 'heave'. Nothing to do with the sickness of the crew; it is simply the up-and-down motion of the ship in response to the swell in the open ocean. This heave is transmitted to the derrick, which transmits it to the drill string and it can all end up with the drilling bit crashing up and down in the hole at the bottom of the sea-bed. At the very least it

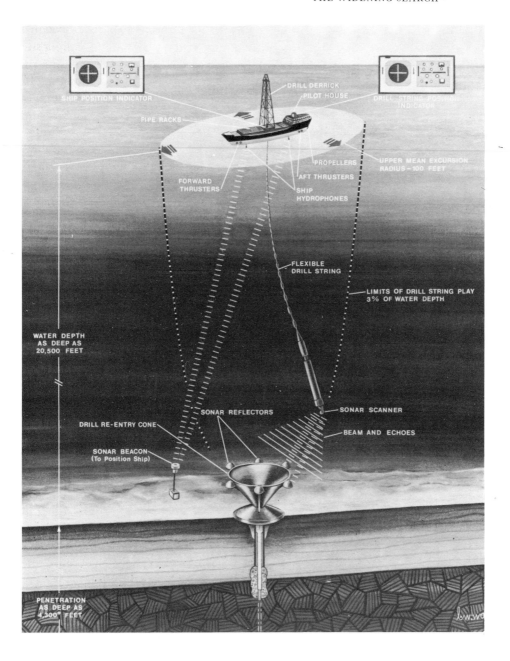

The technique for re-entering the drilling hole that has proved critical to the success of the Deep Sea Drilling Project. While the vessel remains locked in position on the surface, the drill string operator slowly lowers the drilling bit towards the re-entry cone, steering it with the aid of the sonar scanner

can mean a very disturbed drilling core with all the problems that poses for subsequent scientific analysis. At worst it can mean the loss of the whole bottom assembly.

A Californian contractor came up with the answer: a heave compensation unit that effectively insulates the drilling string from the movement of the ship in a swell. They have introduced a massively strong steel cylinder containing a piston between the drive unit and the drill string itself. The upper surface of the piston is at normal atmospheric pressure. The under surface of the piston, and the entire weight of the drill string, are supported by hydraulic fluid at very high pressure.

So when a wave reaches the drillship it moves the ship and the steel cylinder upwards. Immediately a sensing device on the cylinder detects the movements and pumps fluid out from under the piston. So the piston, the drill string and most important, the drilling bit, remain virtually stationary. When the wave trough passes and the ship starts to sink, more hydraulic fluid is pumped into the underside of the piston so that again the ship and the cylinder move, but the piston and the drilling bit do not. Result: elimination of damage caused by heave.

All these developments are vital to the oil industry in extending their drilling range out into deeper waters, and already they are at work. Five drillships are probing the sedimentary rocks of the continental slope from the Arctic seas off Canada, to the South China sea off Thailand. They are all in search of the underwater equivalent of a Saudi Arabia or a Kuwait.

But of course, discovering the oil in deep water is only the beginning of a very long story. Up till now, even in the deepest waters of the North Sea, the solution to getting the oil ashore has been to build a giant production platform over 200 metres high standing on the sea-bed and gathering the oil from fifty or more oil wells. The oil is then pumped ashore along a sea-bed pipeline costing $2 million a kilometre to lay, or into a single buoy mooring alongside from which it is pumped directly into oil tankers.

But beyond 200 metres the standing platform is no longer really feasible, and the oil companies are having to go for the far more complex sea-bed completion and production systems. There are two basic types, one manned, the other almost entirely automated. The French manned system, for example, consists of a number of production stations fixed to the sea-bed, collecting the oil from a number of wells grouped around them. The stations then pump the oil to a central storage platform in shallower water, from where it is pumped directly into tankers. Each of the production stations has a sea-bed capsule that contains all the control and operating equipment. When maintenance is needed the service ship lowers an engineer in a diving-bell that docks with the sea-bed capsule. So the engineer works the whole time in a dry 'shirt-sleeves' environment at normal atmospheric pressure.

For deeper water one of the American multinationals has come up with a fully automated system that is at the moment being tested as a prototype. It involves lowering to the sea-bed a large steel 'crate' or template as they call it, that carries the collecting equipment for up to forty oil wells. From there the oil is pumped up to a single point mooring buoy on the surface and thence into tankers. The control systems have all been designed to be maintained by a remotely controlled maintenance manipulator that would be lowered from the service ships and then guided through a series of maintenance operations by engineers working through closed-circuit TV. Only *in extremis* would they attempt to send in a diver or a submersible.

The oil industry is in many ways the marine equivalent of the aerospace industry. Because the stakes are so high it is constantly pushing the engineers to the very limits of technology. Whenever it comes up against a new problem, a new barrier, it focuses enormous energy into cutting a path around it. I remember speaking to an American engineer a year or so ago, now working on the North Sea production platforms. 'We came over here,' he said, 'thinking we knew it all from our experience in the Gulf of Mexico. We hadn't been here a week before we knew we had to develop a whole new technology to get that oil out of the North Sea.' In many ways the exploitation of the continental slope is going to make the North Sea look a bit like a boating trip on the Serpentine.

Man Under the Sea

On 5 December 1962, a young Swiss mathematician, Hannes Keller, and a highly experienced English diver, Peter Small, were sitting cramped into a small diving-bell as it was winched over the side of a ship anchored in the Catalina Straits off Los Angeles. Slowly, very slowly, the bell was lowered into the sea. As the cable was paid out it sank out of sight and the dive controllers began to count off the depth. Fifty feet, 100 feet, 250 feet. The diving-bell was being compressed with a still secret mixture of gases. The target depth was the magic figure of 1000 feet. The deepest Keller had been until this dive was 660 feet. As the diving-bell inched its way down towards the 1000-feet mark, the divers were watched over closed-circuit TV. They reached bottom. They had achieved a remarkable dive. Then suddenly things began to go terribly wrong. Hannes Keller claims that he opened the hatch and made a brief foray into the black icy water. But Peter Small collapsed in the bell. Keller got back in somehow – half-closed the hatch, then he too collapsed. The bell was raised to the surface as fast as the decompression tables allowed. But Peter Small was dead when he reached the surface. Another crewman died in the efforts to get another pressure line attached to the bell at 200 feet.

Hannes Keller's history-making dive illustrates in many ways the key elements in the push to get divers deeper: the enormous courage involved, the risks taken, the suddenness of the disaster, the ignorance about the detailed physiological effects of intense pressure on the human body. Even today this is true. At 300 metres there is a pressure of a quarter of a ton on every square inch of the body.

So why do divers put themselves so much at risk? Driven perhaps by the constant human urge to extend the frontiers of knowledge? Or do they dive ever deeper because of some primaeval urge to conquer everything, to overcome every element? Not really. The short answer has got to be, because of oil.

As the world-wide search for oil has pushed rigs, pipelines and wells out into ever deeper water, so the divers have been carried with them; because divers, like astronauts, carry between their ears by far the most versatile, flexible, capable computer ever designed. The oil companies have moved into the relatively shallow waters of the continental shelf with a modified land-based technology. So they still need men, divers, to carry out hundreds of essential tasks from installing the 'christmas tree' – the assembly of lines and valves that controls the rate of oil flow from an undersea well – to carrying out emergency inspection and maintenance work. But how long can it go on? How much deeper can the diver go?

Strangely enough that turns out to be a question for which it is difficult to get a straight answer. Mainly perhaps because on so many occasions over the past fifteen or sixteen years the supposedly definitive 'depth barrier' has been broken. But there are two clear limitations. One physiological, the other, the sheer cost of getting a man to great depths and then keeping him alive there. For deep water is as alien and hostile to man as outer space. The body is a soft, porous, fragile structure that clearly was not designed for great changes in pressure. The lungs can collapse, or explode. Poisonous gases can be forced into the bloodstream and be carried to the brain, and vulnerable tissues like the brain and the nerve cells are very easily damaged by changes in the pressure or the chemistry of the fluids and tissues around them. None of the major problems encountered by deep diving has been entirely resolved.

Nitrogen Narcosis This occurs at about 60 or 70 metres when the diver is breathing ordinary compressed air. Nitrogen is, of course, one of the common gases used in hospital anaesthetics. The symptoms vary from diver to diver but generally he is likely to go through a dizzy, drunken, euphoric state, during which he may even see hallucinations before he slips into lethargy and unconsciousness. The remedy has been to replace the nitrogen in the breathing mixture with helium and this, it seems, can be breathed more or less safely down to much greater depths – but it brings its own problems.

It conducts heat out of the body very rapidly. At 200 metres or so it removes as much heat as the body can produce so it has to be heated before the diver breathes it; which means the diver has to carry around a somewhat cumbersome back-pack heater or a heating system on board the mother ship. Helium also produces the well-known 'Donald Duck' voice which has been only partially overcome by hand-held unscrambling devices, not unlike a racing commentator's micro-phone, developed by the Royal Navy.

Heat Loss   Rapid heat loss occurs partly because of low sea temperatures at depth, around 4°C, but mainly because of breathing helium which sucks the heat out of the body. Even the small drop in body temperature of 3°C or so can lead to the muscles becoming rigid, amnesia and complete mental withdrawal. A drop of 5°C can mean death. There are two modern answers to the heat loss problem, both only partially successful. One is electrical underwear. The diver in fact wears four layers. First a cotton undersuit, then an electrically heated woollen garment, then a nylon layer, and finally the rubber diving suit; and still his hands and feet get cold. The second method is the 'hot water bottle'. The diver wears a thick rubber diving suit, and warm water is pumped around his body, escaping out of the finger tips. The problems with this method are that the diver is continuously wet, which can soon become unpleasant; and the supply of the warm water itself.

It has to be pumped down from the surface support ship because that is the only vessel around that has the capacity to carry a big enough boiler, and that means that there can be considerable difficulty in controlling the water temperature at the customer's end of the line. If the divers are operating out of a submersible, either method of heating is a very severe drain on the submersible's limited battery power.

The Helium Barrier   At fast rates of descent below 100 metres, helium can lead to the high pressure nervous syndrome or HPNS; it is thought to be caused by the compression of the nerve cell membranes. It results in violent trembling of the hands and arms or even the whole body, disturbed brain rhythms, dizziness and nausea, loss of balance, and pains in the limbs and joints, often called by divers 'dry joint' because they say it feels as if there is no lubrication in the shoulders and elbows. In the early seventies it was thought that, at a depth of approximately 300 metres, HPNS was the ultimate barrier, but subsequent research in this country has shown that it is possible to push divers past 300 metres provided the rate of descent is slowed right down from around 30 metres a minute to around 3–5 metres a minute, or 120 metres a day. Even at this rate severe tremors and sickness are likely to occur in most divers but they will normally pass off if the descent is a stepped one. That is, if the divers spend long periods, often as much as a day at, say, 150 metres, before dropping down to 300 metres. But of course the longer they take to descend, the longer they are breathing gases at high pressure. That means they will have to spend much longer on the return journey decompressing – getting the pressurised gases out. If divers decompress too rapidly the press-urised gases that have been absorbed literally 'bubble' out as tiny pockets of gas that at least mean severe pain, at worst paralysis and death.

You can see what all this means in practice if you look at the profiles of some of the record-breaking dives. In 1970 two British divers, Bevan and Sharphouse, went down to 457 metres, in a dive simulator – a pressure chamber. It took almost four days to reach that depth and they spent ten hours down there. They then spent eleven days getting back to the surface. Even so, despite the constant care and attention and the slow rates of compression and decompression, they both experienced several long bouts of sickness, dizziness and trembling. In October 1977, Comex, the big French International diving company, filled the headlines with its announcement of a new open sea diving record of 501 metres. Undoubtedly a staggering achievement, but it is worth pointing out that the dives were made by six men who had been hand-picked after a rigorous medical and psychological selection process. They had been through months of special training including two deep acclimatisation dives, and the sea dive itself took fourteen days to accomplish during which the divers were at 500 metres for ten minutes, and at 460 metres for from periods of from thirty-five minutes to two hours. It is impossible to overestimate the courage of the men who carry out that sort of dive, but the result is a phenomenal cost for the oil companies which use them. Divers are very, very expensive. One hour's work at 200 metres means that the diver has to spend around forty hours sitting playing poker in a decompression chamber.

It is directly because of these enormously extended decompression times that 'saturation' diving techniques have been developed. Decompression time depends upon the depth dived, and the length of time spent at depth. Every extra hour 'on bottom' adds many more hours to the return decompression time. But after approximately ten hours at depth the diver reaches 'saturation' point. His tissues are saturated with the compressed gas, they cannot absorb any more. From then on, however long he stays down there, ten hours or ten days for that matter, he requires the same length of decompression time – something of the order of twelve days for a dive to 300 metres. Now in practice, of course, he cannot stay under water at that depth for much more than one or two hours at a time because of cold and fatigue.

But he can remain 'compressed' to that depth in a chamber. The divers spend a day being taken slowly down to the working depth, say 200 metres, in a diving bell. After they have carried out their work schedule they re-enter the bell and are brought back to the surface. But the inside of the diving bell remains pressurised to an equivalent of 200 metres. Then it locks into a larger decompression chamber on the mother vessel which is pressurised to the same depth. This has beds and toilets and washing facilities, radio and TV – and cards. Divers are great card players! They have so much time in which to sharpen their skills. Food is passed through a pressurised hatch so that the divers remain 'saturated' for several days, during which time they will carry out regular working dives of an hour or so, twice every twenty-four hours. The limit to saturation diving is fatigue, because to work repeatedly at depth is exhausting, and boring. The divers say that they may not get the 'bends', but they certainly 'go round the bend' because there is nothing to do but eat and sleep. At the end of the saturation period the diver has a single long decompression and then two weeks' leave before he can dive again.

So even with saturation diving, the cost of diving operations is very high indeed. Any dive below 200 metres has to be handled almost like a space mission. To get even four or five hours work a day at depth requires a large team of skilled divers and it is estimated that their efficiency under water is only about 20 per cent of that on land. It requires properly equipped support ships with one or perhaps two habitable decompression chambers; several thousand pounds worth of helium alone in a week; highly skilled support teams, and complex monitoring equipment to control the pressure and temperature of the gases being fed to the divers. The amount of variation, which is safe, in oxygen pressure and in gas temperature is very small. Two divers in the North Sea, for example, were literally roasted alive because the helium temperature rose too high. Insurance premiums alone on diving operations are now a substantial part of the costs.

Research continues to reduce the dangers and improve the safety factors, but it's a bit like devising a safety code for wall of death riders. Diving is a risky business. And research continues to try to find ways of cutting the costs. The French, for example, have perfected a technique called 'excursion diving' to cut decompression times. If work has to be carried out at, say, 300 metres, the diver is saturated – that is he lives and sleeps in the decompression chamber – at a slightly shallower depth. The difference is made up in the diving bell. It is pressurised each day to the lower working depth on descent, and depressurised again to the living depth on ascent. The overall effect is to lop several days off the final decompression.

In America, a great British diving expert, Dr Peter Bennett, lost to Britain essentially through lack of research funds, has pioneered the reintroduction of a small amount of nitrogen into the breathing mixture. The theory is that since the high-pressure nervous syndrome is thought to be produced by a compression of the cell membranes, leading to hyper-excitability of the nervous system, nitrogen would counteract the effect since it is thought to bring about its anaesthetic effects by swelling the cell membranes. Thus the two effects should cancel each other out. It has been tried out on mice with good results, and human divers breathing the mixture have descended to 300 metres in just thirty-three minutes, and decompressed in four days instead of twelve. But the technique is still far from established.

The ultimate depth for human divers? Well, experts are loath to commit themselves, but it would appear to be at about 600 metres. At that depth it seems that the density of the gases alone will make it difficult for the divers' chest muscles to move the gas in and out, and the high-pressure nervous syndrome is likely to occur to some extent in any diver who remains for any length of time at that depth. But it is clear that, well before that depth, the cost and the very real working limitations of the diver at depth make other systems an imperative.

Jim

Like Jim, for example. Jim is something of a character. He is well over 6 feet tall. He weighs nearly 500 kg. And he has been pressure tested to specifications laid down by Lloyd's Register of Shipping. He is also something of a survivor. He was first developed in the early 1930s. He was laid to rest during the war, and only in the past decade or so has he been resurrected. Jim is an armoured diving suit. If you look at him propped up against his specially designed stand he seems not outdated, but clumsy. Massive rounded body topped by a huge unwieldy dome of a head. Watch him move around under water and although he has some of the stiffness of the classic hard-hat diver, he is amazingly agile, and most important of all, dextrous. As one of the divers put it, 'you can't exactly assemble a watch with these pincers, but then you don't normally assemble watches under water'. But you can bolt on a pipe flange, or turn on a valve, or hold a drill, or an oxyacetylene cutter. In fact, you can do most of the things that a diver has to do, in greater safety, because Jim is a one-atmosphere suit. You can also work in greater comfort, and at greater depths. Jim is at present rated to 540 metres although the system has been pressure tested to 600 metres. Several 'sons of Jim' are being developed, some lighter to work at shallower depths, others tougher to go deeper.

The main sections are made of magnesium alloy because of its very high strength-to-weight ratio. Magnesium corrodes very rapidly in sea water, but the designers have accepted that risk and gone for a tough surface sealing. The key to the whole design however lies in the joints. In the early days in the thirties the joints leaked at 60 feet. With advanced design and modern materials they are watertight and they allow the diver a high degree of mobility in the arms and the legs. The diver gets in through the hinged dome at the top. When that is sealed into place he sees through a large porthole in front of the head and one on each side. Jim is then lifted on a crane, or an 'A' frame and lowered into the sea. The suit carries enough oxygen for work missions of about four hours, with a safety reserve of eight to sixteen hours. The diver breathes through a mouth mask that also has the microphone for diver-to-ship communications and Jim carries emergency through-water sonar systems in case the primary system should fail.

The suit's buoyancy is controlled by a number of external weights and these are adjusted to suit the type of sea-bed. In an emergency the weights and the lifting cable can be jettisoned and Jim floats to the surface. Since the diver inside is

Jim *gets to work with his manipulators on a tricky piece of sea-bed assembly work*

always at normal atmospheric pressure, there are no decompression problems. He can come up from 300 metres in five minutes. And that, in many ways, is the greatest advantage of them all. The diver is released from the tyranny of the compression and decompression tables, and from the cold, and the constant mind-numbing threat of equipment failure at depth.

The diver can be an engineer or a surveyor who has had a very short diving course, something less than fifty hours because that, it's claimed, is all that it takes to master Jim in all but the most difficult conditions. His biggest troubles are deep-sea-bed mud, or clusters of large rocks.

For the oil companies the economics of armoured-suit diving would seem, on the face of it, to offer several attractions; last year, for example, a fifty-seven-year-old diver spent six hours working on the bottom in the Arctic seas off Canada, at a depth of 300 metres. There is no need for a special support ship. The suit can be slung from any vessel with a small crane. A team of three engineers and three divers, with an armoured suit, could, theoretically, deliver over twenty hours at depth out of every twenty-four.

There is no suggestion as yet that Jim could entirely replace free divers, but there's no doubt that in many situations, he offers an armoured fistful of advantages.

Jim does not perform very well when he is left dangling in the water. He is definitely a sea-bed performer. But he has recently been joined by a rather more effete-looking cousin whose sphere of operations is specifically the mid-water zone – Wasp.

Wasp

Wasp is in a sense a mini, one-man submersible, designed independently by a brilliant young British diver. Wasp was first shown in prototype at the Houston Marine Technology Conference a year or so ago and was immediately snapped up by the off-shore industry. It is 2.1 metres long and weighs 500 kg in air. It is made of cast aluminium and glass-reinforced plastic and is rated down to 610 metres; and, as with Jim, the diver remains throughout at normal atmospheric pressure. Power, air, and communications are provided through an armoured umbilical cable to the surface ship. Wasp gets its name from its appearance: black and yellow body-shell and tiny fluttering wings at the waist – because Wasp is controlled by means of its 'flying' motors.

Cowled fans mounted on each side can be rotated about their axes to give thrust through 360°. They are operated by the diver through two foot controls and he literally flies the suit through the water. In an emergency all the external gear, and the umbilical cable, can be dropped and the Wasp can free swim to the surface. Batteries give him forty minutes of emergency power and it carries air bottles for thirty-six hours.

The arms are completely articulated and are very easy to move in any direction. In the event of hydraulic fluid loss at the joints they seal on to secondary seals so that no water gets in.

The key advantages of Wasp are similar to those of Jim: the freedom from all the complex support systems and the physiological problems associated with saturation diving. Wasp undoubtedly has its own limitations – it would soon be in

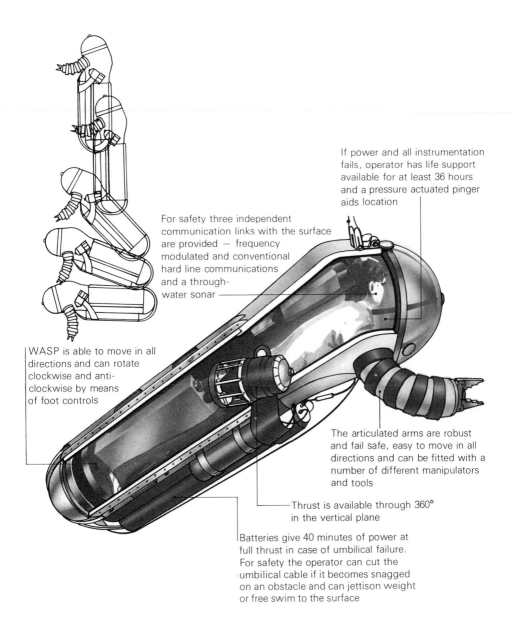

If power and all instrumentation fails, operator has life support available for at least 36 hours and a pressure actuated pinger aids location

For safety three independent communication links with the surface are provided — frequency modulated and conventional hard line communications and a through-water sonar

WASP is able to move in all directions and can rotate clockwise and anti-clockwise by means of foot controls

The articulated arms are robust and fail safe, easy to move in all directions and can be fitted with a number of different manipulators and tools

Thrust is available through 360° in the vertical plane

Batteries give 40 minutes of power at full thrust in case of umbilical failure. For safety the operator can cut the umbilical cable if it becomes snagged on an obstacle and can jettison weight or free swim to the surface difficulty, for example, operating in strong currents, or in very confined space, but I have actually watched Jim and Wasp go through something of a courtship ritual underwater, and I must say they make a very complementary couple!

Submersibles

Submersibles have come a very long way since Jacques Cousteau launched his small two-man saucer-shaped 'Soucoupe' in 1959. The sixties were in some ways the golden age of oceanography because they gave birth to some of the wildest and most romantic ideas about man living under the sea in great sea-bed pleasure domes, and whizzing around in custom-built submersibles of every shape and size. Throughout the sixties there was a vast amount of experimentation with design, with materials – alloys, steels, titanium, glass, even concrete. But all of the submersibles built had very limited range, endurance, depth, or navigational ability.

Even as recently as 1974, for example, the most common way for a commercial submersible to find its way around was to have a line attached to a tell-tale buoy on the surface to give away its movements so that it could be directed by a mother ship. During the late seventies there have been three or four main lines of development.

There has been an enormous increase in scope and power. Today's commercial submersible weighs up to 20 tonnes, can dive to 1000 metres and range twelve to sixteen kilometres over the sea-bed. With two crews they can, in most circumstances, achieve two four- or five-hour dives in a day. So for many types of work they have a much greater capacity than free divers.

Recently there has also been a far greater emphasis on functionalism. The day of the general submersible – with a few 'poke and look' instruments on the front end – is over. They are far too expensive to build and to run. We are now in the era of the specific craft designed to do a particular job with as few compromises as possible.

Navigation and positioning equipment is now far more sophisticated. The submersible sets out a cluster of sonar beacons on the sea-bed around the work site, and then interrogates them with a transceiver. Thus it gets a series of simultaneous bearings to fix its position precisely. Last year one submersible operating in the North Sea used for the first time an inertial navigation system developed by Ferranti; it was essentially a scaled-down, simplified version of the navigation systems used by high-flying jet aircraft and big nuclear submarines. It consists of a series of interlinked gyroscopes to measure every acceleration, every change of direction from a given starting point. All the information is fed into a small computer and it continuously plots the submarine's position. In this case it is claimed that it gave detailed measurements to within 20 centimetres. Nineteen

A *NR 1 – Nuclear Powered Research submersible* B *Seacliff can dive over 1000 metres* C *CURV III – Cable Controlled Unmanned Recovery Vehicle* D *DSRV1 – Deep Submergence Rescue Vehicle*

years ago the 'Soucoupe' made its dives with just a compass!

But even with the best navigation equipment in the world, the submersible still has trouble seeing where it's going. It is dark down there. Flying a submersible around the legs of some giant sea-bed platform or along a pipeline 300 metres down has been described as like driving a car in fog at night across a field. Even high-powered searchlights do not do much to lighten the darkness. But sound does. High-powered, high-frequency sonar scanners are the eyes and ears of the submersible. Forward-looking sonar picks up obstacles; the outline of a wrecked ship, a rock face or a sea-bed oilwell. Downward-looking sonar gives a constant record of height above the sea-bed, and side-scan gives a continuous print-out of the topography of the sea-bed. Armed with this sort of equipment and a range of specialised tools, manned submersibles now dig trenches, bury cables, repair pipelines and oil-well heads, and do a hundred and one different kinds of survey and sampling work.

Perhaps the biggest area of development over the past two or three years is the diver lockout submarine like the latest of the Vickers LR Class of Submersibles that was launched last year. Basically, it is a submarine with two separate compartments. The first one carries the crew of two pilots and all the control gear and it is maintained at normal atmospheric pressure.

The rear compartment carries two divers and is pressurised to the depth at which the work is going to be carried out. So, when the submersible reaches the work site the divers can leave the hull through an air-tight hatch, carry out the mission and then return to their pressurised chamber. One of the crew is often an engineer so that he can give the divers precise instructions by watching them through the observation panel as they are carrying out their work. It is not a new idea; it was first used in World War II to put diver commandos into the sea close to their target. But with its modern development, in lighter, far more manoeuvrable submersibles, it means that a team of divers can be carried around a sea-bed oilfield, making repeated excursions to carry out inspections or maintenance work, before being carried back to the surface.

One of the most remarkable things about the LR Class is that the pressure hulls are capable of withstanding pressures down to 500 metres and yet they are made of plastic, or glass reinforced plastic (GRP). The weight of the pressure hull is now a major factor in the design of any submersible, since, as with aircraft, the lighter the hull, the greater the payload that can be carried for the same power. GRP was used only after several prototype hulls had been tested to destruction; quite apart from its lightness it is said to bring other advantages with it as well, such as less in-hull condensation, and greater warmth.

But manned submersibles of every variety still have major limitations. They are slow, so they have trouble with strong currents. Their range is still severely limited by their battery power supply. They are unwieldy in the sense that they cannot manoeuvre inside the legs of a platform, for example. But above all, they are immensely costly to operate. Those are the main reasons for the fact that the fastest-growing area of the submarine business is now the unmanned submersible.

In general they look crab-like and ungainly. They have none of the smooth, streamlined appearance that we expect of a submarine. But the unmanned craft can carry very much the same range of TV camera, sonar scanner, grabs, manipulators and drills as its manned counterpart and can tackle many of the same jobs. It is controlled by an operator on board the surface vessel through an umbilical cord that carries power and commands to the submersible, and a stream of data back to the recorders on the control ship. It is of course much cheaper to design because it does not have to carry life-support systems for any crew, it can be launched from any surface ship and it can work at very economic rates – round the clock if necessary.

The biggest market for unmanned submersibles is undoubtedly in the bread-and-butter work of pipeline and rig inspection and sea-bed survey work and there is no doubt about their usefulness in a tight corner. It was the utterly functional-looking CURV III (Cable Controlled Unmanned Recovery Vehicle) which belongs to the US Navy that finally got down to 500 metres in the Irish Sea and attached a cable to the stranded British Pisces submersible so that it, and its two-man crew, could be winched to the surface. In fact search and rescue is still one of the great weaknesses of the industry. Despite the growing number of submersibles

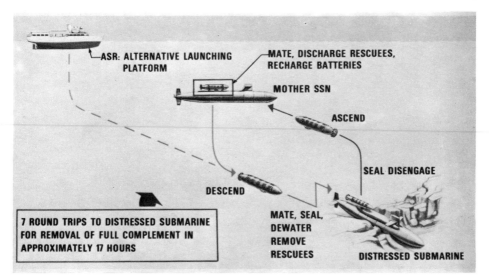

Possible profile of deep sea rescue mission using the DSRV

throughout the world, there is no coordinated rescue organisation or system. Nothing to resemble, on a commercial scale, the American Navy's Deep Submergence Rescue Vehicle (DSRV1) for example, which has three pressure cabins to enable it to make repeated dives to a nuclear submarine in trouble on the seabed and ferry passengers to either a surface rescue ship or a mother submarine. DSRV1 is now backed up by the Deep Submergence Vehicle, Sea Cliff, that was used last year at depths of over 1000 metres to recover parts of a top-secret Navy Tomcat fighter aircraft that had plunged into the sea ten miles off the Mexican coast.

What of the Future?

The two main areas of development are going to be in power units and in materials, and in both America holds a clear lead. The American Navy's NR1 is the world's first nuclear-powered submersible. She is described officially as a research vessel although there are very few details released about either the nature of the research or the design, except for the radical feature of her main drive motors being positioned outside the hull. The NR1 carries the world's smallest nuclear reactor, a crew of five, and it can stay deep under water for weeks on end. It could prove to be a turning point in submersible design.

An even more radical approach to submarine drive is so-called 'buoyancy propulsion'. For all but local manoeuvring the submarine would simply harness the effects of negative and positive buoyancy. Its only fuel would be buoyancy gas. Its only 'drive' machinery would be short stubby wings or lifting surfaces, and control vanes. So, it would dive from the surface under the weight of its filled ballast tanks until it reached its limiting or working depth, then it would simply empty its ballast tanks and 'soar' to the surface again driven by its positive buoyancy. It is estimated that a military-sized submarine could reach speeds of fifty knots and travel more than sixty miles in a single dive/climb sequence. The feasibility of the idea was demonstrated as long ago as 1957. So far no one has taken it up.

As far as materials go, the front runners for the future are certainly acrylic and glass. Again the American Navy pioneered the use of acrylic with its NEMO observation sphere and it has since been taken up in an American working submersible, the Johnson Sea Link, which has a completely transparent acrylic two-man pressure hull. Apart from the obvious advantages in visibility – it is like going down inside a goldfish bowl – acrylic offers the almost perfect combination of lightness and strength, down to about 300 metres. Below that depth it has been argued that the perfect material would be glass, because it becomes practically indestructible under high pressure. There are already one or two submersibles with hemispherical glass viewing chambers, but as yet no one has made a complete vessel because no one has been able to make large glass spheres good enough; any flaw would provide a weak point that might collapse under pressure. An even more revolutionary submarine building material is called syntactic foam, made of plastic resin and minute glass spheres; it's remarkably tough and light.

So, the deep-diving submersible of the year 2000 could well be a dazzlingly beautiful structure of glass and scintillating plastic foam, gliding silently on its wings down to the deep-sea oil fields on the edge of the continental margin. It looks as if the 1980s will be the decade of 'inner space', the decade of enormously expanded ocean exploitation.

Michael Rodd, Michael Blakstad

Telecommunications

part one: The Hardware

It is hard to imagine what life must have been like when there was no telephone, no telex, no television. When the only means of communication was the spoken or written word, and when carrier pigeons were possibly the fastest way of getting a message from one place to another.

Our forefathers survived and no doubt enjoyed a less hectic lifestyle than the one all these technological marvels have given us, but it would be exceedingly difficult for us to return to those days. The way we live now depends on fast, accurate communication sometimes over great distances.

Apart from the spoken and the written word, we have two other allies in our fight to communicate better. One is the radio wave, and the other is an electric current through a metal conductor, communication by wire. Between them these two advances have brought mankind a long way, and will take us a great deal further.

Telephones and Airwaves

One of the prime reasons for building London's Post Office Tower was to lift telephone calls almost 180 metres above the cluttered roof tops and beam them on their way as microwave signals. All over the country even the least observant of us can spot other links in the chain. The towers are usually on hilltops, supporting the directional dishes which pick up the microwaves and retransmit them in the exact direction of the next tower.

Microwaves present advantages over the conventional means of transmitting telephone conversations down a pair of copper cables. For a start you do not need the cables themselves so there are no unsightly pylons to build or expensive holes to dig if the cables are to be buried underground. There are, of course, the towers themselves to build which are quite substantial structures and have to be positioned no more than twenty-five miles apart.

Microwave transmission adopts a technique called frequency division multiplexing (FDM). It allows more than one conversation to be carried at the same time through each channel. Briefly, the waveform, the line which represents the rise and fall in the sounds of the conversation, is passed through a multiplexer. This gives each waveform an artificial centre line, each one different from the others. In other words, the frequency range of each conversation (and on the telephone it is limited to 3 kilohertz by the equipment anyway) is given a specific region of a much wider frequency band. It is almost like giving each conversation a different musical key. At the other end the different frequency bands identify the different conversations which, once separated, are converted back to the original frequency band. At either end they sound perfectly normal despite their key change *en route*.

Microwave transmissions give each conversation a 4-kilohertz band to allow safe space between different callers, and transmissions are made in the frequency range between 5 and 6 gigahertz. That is between 5 and 6 million kilohertz, so you can get some idea of the key changes that take place!

By collecting the channels into groups, each group can then be multiplexed again into a supergroup carrying just under 1000 different telephone conversations. And good though that may sound, engineers are constantly struggling to meet our growing demand for communication channels.

Microwave transmissions also have some disadvantages. The line of sight that is needed between the towers can be a problem and the expense of building the towers themselves partially balances the savings of not having to install cables the entire length of the connection. All radio transmissions are also susceptible to adverse weather and interference of various kinds. Engineers have been working for some thirty years or more to develop a system for using the advantages of microwaves, whilst increasing their capacity and reducing the danger of interference and interruption.

Bending the Airwaves

The answer seems to be the waveguide. In principle it is a combination of airwave and cable communication. The microwave signals are fed into a copper-lined tube which contains the signals, forces them to turn corners and prevents them from picking up any interference. Safe in their copper tube the transmission frequencies can be stepped up to between 30 and 110 gigahertz.

Remembering that the old, fresh-air microwave transmissions manage to squeeze 1000 simultaneous calls between 5 and 6 gigahertz, it does not take a mathematical genius to work out that between 30 and 110 gigahertz, a total of 80,000 should be possible. In fact by introducing another multiplex system on top of FDM an amazing half million calls can be squeezed in.

That is the sort of capacity which today's communication engineers find attractive. The system on which it depends uses pulse code modulation. This involves converting the analogue waveforms, the lines of the FDM signal, into digital pulses.

Having grasped the idea of pulses representing curves, it is not difficult to accept that there are holes in the pulse picture. Like a graph plotted from point to

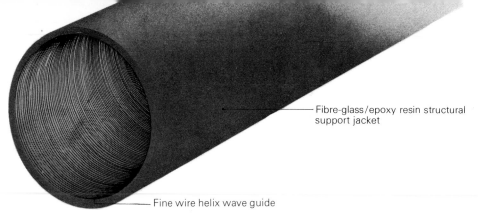

Fibre-glass/epoxy resin structural support jacket

Fine wire helix wave guide

Waveguide

point, comparatively few points can produce an extremely precise curve. In fact, a curve sample taken 8000 times a second is enough to reconstruct an accurate analogue voice curve later on, and it leaves plenty of room between samples to put information in concerning a totally different curve. This type of multiplexing is called 'time division multiplexing' and the number of unrelated curves that can be carried depends simply on how many fractions the equipment can break each second into.

Digital equipment can perform so quickly that by combining time multiplexed, pulse code modulated signals with the frequency division multiplexing, waveguides promise to carry almost 500,000 two-way conversations at the same time. They need repeaters every twenty miles, but since the Americans conceived the idea in 1938, the slight disadvantages seem to have been outweighed by the substantial advantages that these amazing copper tubes offered.

But you and I are unlikely ever to make a telephone call along a copper tube, because in March 1978 the decision was made to forget about waveguide at least as far as Britain is concerned, and to write off the R and D costs. These will amount to more than five, and less than ten, million pounds. 'Why?' you ask.

Waveguide has proved to have a high signal carrying capacity, but they are expensive beasts. That expense can only be justified if the capacity is used to the full, and all the indications are that it will not be. Demand for circuits has not grown as fast as was predicted and there is no point in investing in capacity that no one intends to use.

Making a waveguide is a precision job and laying one is more complicated than putting down any ordinary cable. First a 100 mm steel tube is buried, then the 50 mm waveguide has to be fed into the centre of this protective outer layer. No one knows what it would cost, because no one has ever done it commercially, but it would take a whole week to lay just 1 kilometre.

Even the Japanese, who have made the running recently in so many other areas of electronic technology, seem less than anxious to take advantage of everyone else's reluctance to get into waveguide. The reason is the emergence of a more simple and more versatile means of communication – optical fibres.

Cables of Glass

It is an old idea. John Logie Baird was just one early twentieth-century pioneer who recognised that light could be captured inside a piece of flexible glass and transmitted some distance. He failed to find either the right glass or the right light, but no such problem confronts today's communications engineers.

The glass contains the light rather than disperses it, because the rays of light are reflected off the inner surface of the glass, and not transmitted straight through. One technique is to produce a glass rod with one type of glass in the centre surrounded by glass of a lower refractive index. This means that light in the centre section can never break out through the outer layer. It is doomed to bounce along the inner core, zigzagging as it is reflected off the inside skin of the outer layer.

All the world's major glass and communication equipment makers are experimenting with optical fibre production, and such is the commercial prize that the patent lawyers are already enjoying something of a boom defending their particular clients' interests and attacking all who come dangerously close.

Once the rod is heated and the fibre has been drawn off, the glass is no thicker than a human hair. It is usually protected in a plastic coating and ends up wound on a drum not unlike the copper cable it may one day replace.

The light source is that now familiar breakthrough of the 1960s – the laser. Usually it is a semiconductor laser in which the atoms of the semiconducting material are excited and made to lase by the electric current. Anyone in doubt should refer to the chapter on leisure for a fuller explanation of the mysteries of the laser. Light-emitting diodes are another possibility for the light source, and these would be cheaper to install than lasers. They do not produce such a fine pencil of light as a laser, however, and would only be suitable for installations with comparatively little capacity.

The optical signal would be the light source flashing on and off, each flash representing a pulse of our digital code. All the principles learned from waveguides are of value to our understanding of how optical fibres work. Unlike waveguides, optical carriers will be simple to lay. They can lie in the same duct as conventional cables, and can often be squeezed in when there would not be room for another copper wire.

Tomorrow's World has reported on the techniques developed for joining lengths of cable together. Unlike a copper electric conductor, it is not enough for the two cables to touch. The central cores of the light-carrying glass must be lined up precisely, otherwise the path of the light pulses is interrupted and the signal does not cross the joint. The techniques for ensuring perfect connections have been developed and the indications are that a glass will eventually be developed which is pure enough to transmit over thirty-five miles without a repeater.

British manufacturers of telephone repeater stations are not too enthusiastic

A few optical cables offer the same capacity as the copper cable ducts on the right

about a reduction in their business but the savings to all of us should be substantial. Even if all that is still some time in the future, optical cables already promise to carry 10,000 simultaneous telephone conversations in much less space than that taken today by a copper cable which can only manage 5000.

Work also goes on developing a photo diode detector for the receiving end of the circuit. At the moment the Canadians make the ones under test with the British Post Office. Eventually optical fibres will provide the sort of capacity which waveguide promised, without the expense, and that will leave the telephone authorities with cheap capacity which they can sell to their customers. And that will mean in turn our telephones will become exceedingly versatile communications centres.

Tomorrow's Phone Calls

One day we all may find it useful to have a facility for sending documents, writing and pictures, across the telephone lines. A detector at the sending end would quickly transmit the signals representing the document to a printer at the receiving end where the document would be accurately and quickly reproduced.

View-phone would become an easy facility to provide; not that the findings in

America indicate an overwhelming demand for us all to be seen as well as heard on the telephone.

Conference telephone facilities could become widely available and multi-channel cable television could also consume some of that capacity, but perhaps the greatest use, initially at least, will be in what our Post Office now calls Prestel.

The system will link the subscriber's television set to a public computer via his telephone. By dialling the appropriate number on the telephone and connecting the television receiver to the circuit, the user will have at his disposal sixty thousand or so pages of information, and all sorts of services which the telephone and the television on their own could never provide.

The television receiver is fitted with a small memory capable of storing the digital information necessary to generate one page of display. The conversion to Prestel also requires fitting the television with a key pad, rather like the finger panel of a pocket calculator.

On dialling the central computer the massive memory there transmits the signals to the small memory on the television set. These signals generate the first page of the instruction sequence. The picture welcomes you to Prestel and gives you any message it has stored for you in its memory. You, of course, can leave a message for someone else to pick up. Prestel talks to you in that strange computer fashion of asking you questions to which you can answer, 'Yes', 'No', or give a number.

By asking questions the Prestel computer finds out what services you require, whether you want to find out the closing prices on the Stock Exchange, what is on at the theatre or something more complicated. Your answers on your key pad dictate when the memory in your set will receive a new set of instructions from central control and what those instructions will be.

Leaving a complicated message may prove difficult but as long as you are prepared to accept the alternatives Prestel offers, your wife can, for instance, learn whilst you are on your way, that you are arriving at the station at the time you have keyed in from your number pad, or that you are not coming home at all.

Then there are more subtle applications. On file in the main memory could be a whole host of valuable data ranging from, for instance, the Highway Code, to when you are likely to be approved for a mortgage. Because the computer can respond to your 'Yes', 'No', or number answers, it can actually give you advice.

One programme I tried was designed to help those anxious to adopt a child. The first question was, 'Are you applying on behalf of yourself only? If so key zero, if not key one.' I keyed zero, for the sake of argument, and up came question number two. 'Are you the parent of the child?' I keyed zero again indicating I was not. 'Are you related to the child?' Again zero for no. 'Are you over 25?' I lied a little and said 'no' again and this proved too much for the computer. 'You are not eligible to adopt,' came back the answer.

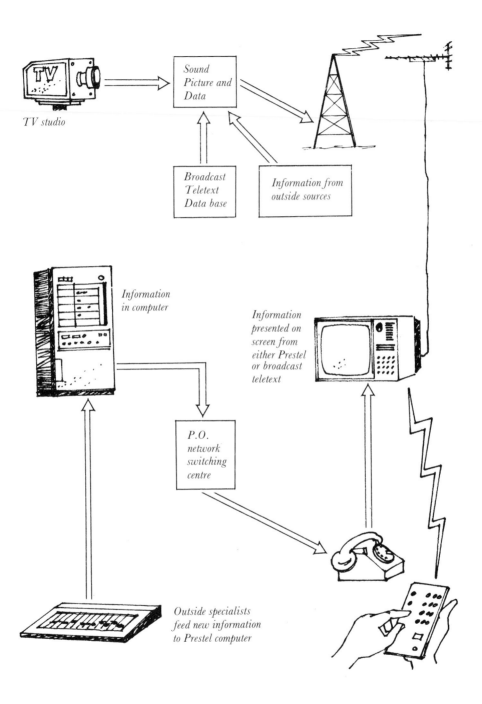

TV studio

Sound Picture and Data

Broadcast Teletext Data base

Information from outside sources

Information in computer

Information presented on screen from either Prestel or broadcast teletext

P.O. network switching centre

Outside specialists feed new information to Prestel computer

If I had told the truth about being over 25 that would have made a difference and another question might have emerged before a final answer was given. Exactly the same technique is being used in some hospitals now for routine diagnosis of patients' complaints. The computers in this case are not public ones on an open network, but maybe the time will come when the Prestel computer will also tell you what is the matter with you, even if it takes a while longer for it actually to prescribe a treatment.

Prestel will also play games with you. Computerised noughts and crosses, mazes, and problems like balancing fuel consumption in the retro-rockets of your Mars lander against the gravitational pull of the planet so that you land at a comfortable speed without running out of fuel, are all in the compendium at computer headquarters.

In fact it is easy to forget that what Prestel is really trying to do is get you to use the phone. The Post Office does not want to engage in games or even take over completely from the Citizen's Advice Bureau, but it is in business to sell telephone calls, and the advent of Prestel gives it a mighty potent marketing weapon. But Prestel and all the other new services which will emerge in its wake can only work if the capacity for these extra services is built into the system. It is optical fibre communication which promises to make that possible.

Just a word perhaps about another sort of teletext which is the generic jargon for Prestel-type displays on television sets. Ceefax, the BBC's teletext system, and Oracle which is put out by Independent Television, both look very similar to Prestel but serve different purposes.

Broadcast teletext is information transmitted in the spare lines at the extreme top of the television picture.

Inside the set is a decoder which extracts the information and displays it as an independent picture on the screen. The viewer is provided again with a key pad but this goes no further than his receiver's memory. Whichever page of information is chosen is simply retained the next time it is transmitted and then it is displayed on the screen. The viewer is not in direct contact with the originating computer and so Ceefax and Oracle cannot ask questions of you, can offer no individual advice nor play games.

But they do provide a right-up-to-the-minute news and information service, and a wide variety of useful data. Broadcast teletext is aimed at everyone; subtleties of direct subscriber-to-computer connection are the province of Prestel.

What is interesting is that both developments are British. Reaction overseas has been mixed. Some American television networks are confused about Ceefax and Oracle. As one executive put it, 'Why give the viewer something else to watch when the commercials are on?'

Whether the Post Office will win public approval for Prestel remains to be seen, but it is interesting that market trials of the technique are being launched in

America even before they get fully under way over here. Everyone is looking for profitable ways to mop up that telecommunications extra capacity.

part two: Is there anybody there?

The principal reason why the telecommunications revolution will happen is that it will make perfectly straightforward use of three technologies which already exist in abundance. Along with the increased capacity fibre optics will produce, there are £500 million worth of copper cables already under the ground as well as microwave circuits linking Britain's telephones together; then there are the television receivers which most homes possess; and finally there are computers which exist in plenty and which are becoming ever smaller, cheaper, and easier to link to the telephone system. It is the computer which will act as the brain and the information store, controlling the switching of the telecommunications network and providing libraries of facts to which the users will have access via their telephones. The TV screens will serve to display the information which then arrives in the home in a form which mere humans can understand, while home keyboards will make it possible to reply to the computers or to other users, and to ask fresh questions. If this chain of linked computers is both soundly conceived and technically well executed, then life will be well worth living in the telecommunications era. If, on the other hand, the system is irresponsibly conceived or simply not very well executed, life could well become an Orwellian nightmare.

Education

For nearly a decade, children in the Twin Cities area – Minneapolis and St Paul – in one of the USA's most northern states have been partly educated by computers. Today, 92 per cent of the school population of the state of Minnesota is educated on computers, 2200 terminals reaching not only to the Twin Cities, but also into the smallest farming communities. The state grants a million dollars a year for the phone bills alone so that the students can communicate with their computers.

The computers do not take over from teachers, of course. Perhaps they could, but no one at the moment thinks that this is a particularly good idea; the computers simply help out. The prize advantage of a computer terminal in a classroom is the fact that it can work just as fast, or as slowly as the individual pupil. However complicated the student may find the mathematics lesson, the computer bides its time; having posed the question it waits for the answer, getting on with the business of checking round all the other 2199 terminals in a continuous and very rapid series of interrogations – Have you answered yet? Have you answered yet? – and when each pupil does respond to the question, the computer checks instantly whether the answer is correct, then calmly tells the child to have another go if it is not, or moves with flickering words of congratulations on to the next programmed problem (always, but always, using the pupil's first name). All the time, the child's performance is being monitored and stored; examinations are obsolete in schools where the computer knows every nuance of a pupil's performance.

The computer does not have to be a completely passive tool in education. It can bring to the classroom excitement of a kind which a mere mortal teacher cannot offer. Those Twin Cities computers can lay on a fully detailed simulation of ecological life in a nearby lake which is so precise that the students can observe the wonders of nature without ever having to step outside into the sub-zero temperatures of the Minnesota winter. Young farmers in this heavily agricultural state can be given whole farms to cultivate, with the results of their actions (allowances made for freaky weather or movements in price thrown in as part of the 'noise' of real life) fed back to them in minute and accurate detail; however bad their performance, they won't have harmed a single grain of wheat. In the same way, would-be sociologists can tinker with a full-scale presidential election in a matter of hours rather than years, with all the unpredictability of the real thing built into the computer model.

To the older observer, one of the most impressive sights at a computer-aided school is to witness six- and seven-year-olds issuing instructions to a computer and reacting to the replies, understanding binary functions as though there were no other way of communicating numbers, bashing the keys and talking to the computer as though it were a favourite uncle.

When the students come to leave school, the computer opens up the fascinating prospect of career guidance. In most British schools, the careers master is a hard-pressed individual with not nearly enough time to get to know either his students or the job market to the extent that is necessary to help children with one of the most important decisions in their lives. The computer, of course, already has a full knowledge of the pupil's performance at school, and it is not difficult for the student to key in a summary of his or her hobbies and special interests, as well as a short list of jobs which, ideally, would suit the candidate. What is now needed – and what is being pioneered by Cheshire County Council – is a data base of the jobs which the authorities think will become available in the years to come. The computer then attempts to match the qualifications and job aspirations on the one hand against careers on the other, and in theory a match can be made. In theory.

In fact, of course, it is, and will always be, impossible to quantify all the human factors needed for a particular career – how does anyone measure the ability to get on with a choleric boss or a militant shop steward? – and it will require an army of civil servants to tabulate all the openings which are likely to occur up and down the country. They could, of course, themselves be helped by computers, but at some point it will surely be noticed just how political the subjects of education and employment really are, and who can tell what the result would be of the inevitable and utterly necessary debate?

Money

IBM executives who arrive at the company's International Training Centre at La Hulpe in Belgium are handed a plastic card, very much like a standard credit card, which introduces them not to a bank but to the building's IBM 3750 telephone exchange. Whenever they make a phone call, they enter the card into a slot built into the telephone and it records precisely how much their call has cost. When they go to the canteen, they plug the same card into the cash register and it tells the computerised telephone exchange just how much to bill them for their food. It gives them a complete cash card service; in fact, as one IBMer put it, the only thing it doesn't do is actually pay the bills, which can be a shock when they arrive. It would be possible, even today, to lessen that shock by linking the telephone exchange's computer to his bank's computer and let them handle the whole transaction.

Many a British housewife is now familiar with the sight of tiny striped labels printed on the bottom of a packet or tin on the supermarket shelves. As she takes the goods she has bought and puts them on to the check-out conveyor, the labels are automatically scanned by a tiny laser beam which records on an electronic cash register exactly how much each purchase has cost. It also informs the supermarket's stock control computer which goods need replacing on the shelves and therefore a fresh order from the supplier. The electronic cash register, too, could be linked to the customer's bank and thus debit her account without money changing hands.

Anyone who banks with one of the big four clearing banks in Britain is probably familiar with feeding his or her personal credit card into a large electronic teller built into the wall of the local branch of the bank, and receiving cash, paying in a cheque, transferring funds from one account to another; allowing, in fact, the machine to perform most of the functions for which, in the past, it was necessary to queue for a human teller – and for which the bank has had to pay the high cost of keeping open a large number of local branches.

At bank headquarters, the central services have been computerised for some years now, which is why personal cheques have carried numbers in a style which computers can read. Human operators translate the handwritten details into the same electronic code, and the cheques and other documents are then not only processed at high speed, but the details are fed down phone lines, to the relevant branches. Transactions which occupied both a great deal of clerical effort and an even greater amount of time, as the postal services shifted paper up and down the country, are now completed in no time at all and with no greater human involvement than the keying in of those last, personal, details.

The motives behind the increasing use of 'electronic funds transfer', as it is coming to be called, are at once obvious and yet bewildering. First, there is the increasing cost of keeping local branches open and paying staff to stay there during the hours that other working people can get to a bank. Electronic tellers are helping to stem increases in either wages or mortgages and rents.

Then there is security: figures are bandied about, listing the total losses to shops and supermarkets caused by pilfering from the till, or checking-out goods at dishonest prices to friends and relatives. The security van loaded with cash from the supermarket or the shopkeeper carrying his week's takings from the till to the bank are, it seems, easy targets for professional crime. All this would be eliminated if money 'went electronic'.

Finally, speed of transaction. We should all like to see our debts settled more rapidly (though we're inclined to pay our own bills as slowly as we can). There is fierce debate amongst economists as to whether rapid movement of money has a positive effect on the economy, but bankers, these days, seem more than eager to hurry the whole business along.

That is why SWIFT was born, and it's why SWIFT is causing one or two flutterings in the dovecotes.

> Swift as a shadow, short as any dream,
> Brief as the lightning in the collied night,
> That, in a spleen, unfolds both heaven and earth,
> And ere a man hath power to say 'Behold!'
> The jaws of darkness do devour it up;
> So quick bright things come to confusion.
> (*Midsummer Night's Dream* 1.i.144)

In 1971, a consortium of Western banks set up the Society for Worldwide Interbank Financial Telecommunications (SWIFT) with the simple idea of taking all the sweat out of international money transfers and speeding the whole business up by handing it over to a computer. The system started operating in 1977 and by 1978 had linked up no fewer than 500 banks in Western Europe and North America.

The system works like this. If an importer in Britain, for example, wants to pay an exporter in America for some goods he has bought, he instructs his bank in London to pay the exporter's bank in America. All the London bank has to do is to type in on its SWIFT terminal the address of the bank in America, the name of the man to be paid and the amount of money involved. The message then passes up to the main British collecting point for SWIFT messages (in fact a couple of wardrobe-sized boxes on a trading estate in Edgware), and it's then routed off to the main computers in Brussels and Antwerp. The computers there look at the address of the American bank and send the message off on its way to the American collection point, where it is forwarded on to the right bank, which then makes the payment – theoretically in a fraction of a second. All very simple and a considerable advance on the cumbersome methods used before SWIFT – or it will be when it's working properly; at the moment SWIFT is only handling about 60,000 transactions a day compared with the 300,000 it could carry. Not enough banks are yet tied into the system and there have been problems with the central post office of the system – the computers in Brussels and Antwerp.

At present, SWIFT covers only a part of the world – Western Europe and North America – but in time it should extend on to Tokyo, Hong Kong, Singapore and the Pacific, and then on round the world to the Middle East, finally connecting up again to the European and American banks. (Several Far Eastern banks have branch offices in Europe already connected to SWIFT.) The final link in a world-wide network could convert SWIFT from being a glorified post office, sending and receiving messages, into a very different kind of bird.

Two key (human) ingredients in the present foreign exchange business are the Bank of England and each bank's foreign dealers. The Bank of England's job is to keep an eye on the movement of money into and out of Britain – to support the pound if it sees the need – and to make sure that no bank enters into contracts it won't be able to meet. It does these jobs by demanding reports from every British bank, stating every hour of the trading day how much foreign currency it is holding.

The dealer's job is to buy and sell currency on behalf of the bank and its clients. For instance, a large multinational company with funds in many different currencies needs to switch from one currency to another as their values go up or down. The multinationals tend to rely on banks to do this for them as the dealers there are specialists with very quick reactions and a good network of information at their fingertips. So, if a dealer spots that, as in early 1978, the dollar is being heavily sold, and that the price will go down, he will very likely sell dollars to protect his clients' and his bank's interests – but he will have to move quickly because the market moves very quickly indeed.

Now what will happen when SWIFT goes on-line around the world? The first thing most banks will do will be to connect SWIFT to their own internal computers which – along with the jobs described earlier – handle part of the international exchange business. Now with the help of the computer, the bank could handle foreign currency deals for twenty-four hours a day both because SWIFT will supply an infinitely better communications system than Europe has had with the Far and Middle East in the past, and because the banks' own computers could perhaps be programmed to carry on simple dealings through the night when the dealers themselves have gone home, but when the Far Eastern exchanges are at their most active.

What is worrying some experts is the thought that SWIFT will then hasten the movement of currencies; had the system been operational in 1978, they argue, the decline of the dollar would not have been a year-long affair but a mad scramble to sell, as computers simultaneously moved to offload their depositors' dollars. 'Sell', the computers' instructions would say, and in an instant, hundreds of computers would be trying to find buyers for the endangered currency. Even when bankers were asleep, the computers would have continued doing business; SWIFT will remove the vital buffers built into today's banking system when the money markets have closed down. No longer will transactions be carried out at the discreet, considered pace which gives the national banks an opportunity to step in and support their own currencies and to calm the frenzy of the dealers. SWIFT, in short, would take the breathing space out of trading.

Nonsense, say the bankers. For one thing SWIFT is nothing more than an extension of the present system; more efficient, we hope, and admittedly faster, but not different in kind. Major deals will still need the approval of senior, human members of staff, and they will still have some quaint old-fashioned concepts built in, like 'value dates' – dates on which payments are to be made which will hold back the fervour of the electronic leviathan. Don't worry, banking has always been operated by good men and true, who are too sensible to be caught in a speed trap of this kind and there's absolutely no reason why safeguards can't be built in to prevent trading if one currency is taking too much of a hiding. Leave it to them, say the bankers, and there will be no cause for concern.

In any case, computerised money will not arrive overnight. Banks in America have been geared for some years to provide electronic funds transfer; what is holding them up over there is the fact that banks are organised on a state-wide basis and not nationally. Any suggestion that they might start linking up and forming any kind of 'cartel' triggers the American anti-trust philosophy (the bankers call it paranoia) and the government acts very quickly indeed. The second largest bank in the country, Citicorp, is attempting an interesting and wholly electronic route round this restriction by using the Visa credit card (of which it plans to be the largest issuer) as a means whereby clients on the West Coast can in fact use the Citicorp's East Coast facilities. When 'depositor' slots his Visa card into an electronic teller in a Citicard Banking Center in Forest Hills, he is linked to the East Coast Bank, and is in a position to conduct as many transactions as the computer can handle. There's no doubt that there, at any rate, telecommunications are changing the nature of banking.

In Britain, there is no need to be so devious. Our banks can and do operate nationally and could switch to electronic money in a very short time indeed. The supermarket checkout system could spread to smaller shops and then to homes, petrol stations, even taxis and window cleaners (small portable units could store the transactions till the owner gets to his on-line terminal at the end of his rounds, then they can be fed to the bank in one batch). The banks will certainly become more confident in their computers, letting them communicate more freely up and down the country with less human interference. The depositor could well become so satisfied with the security and convenience of his magnetic cash card that he will be easily sold on these moves towards a cheaper, more efficient banking service. SWIFT will no doubt be moving from success to success without a lot of publicity; the City prefers to play things that way. And then, one day, someone will quietly decide to tie up all the loose ends.

Privacy

Until recently the social security files of the Canadian Government have been handled by a key punch bureau in, of all places, Korea. It is cheaper to keep the information in Seoul and make entries down long-distance telephone links than to pay the salaries of Canadians to maintain the data in their own country.

In March 1978, the *Guardian* revealed that the police computer at Henley was so overloaded that it had resorted to using county council processors to help with what the Home Office referred to as 'routine chores'. What horrified the newspaper was the thought that the country's most sensitive computer files were separated by only a telephone line from some of the least secure terminals in the country.

The worst nightmare offered by the computerised society is the unfettered exchange between data banks of confidential and sensitive information. There are, for instance, 2000 credit rating bureaux in the United States and at least 100 in Britain, which devour personal information about every individual who applies for any kind of credit, be it the purchase of a television on the never-never or opening an account at a fashionable department store. These bureaux, many of them now computerised, dig through every available file for information, accessing court registers, employment records, insurance and health details, even schools and banks if they can get through to them (and it is surprising just how often they can). One register in Britain has 14 million files on different citizens, but that isn't a patch on one American firm which has 100 million dossiers, more than the FBI and the CIA combined.

It is not surprising, then, that for a decade computer experts and civil liberties representatives alike have been agitating for tighter security on computer files, requiring the right for any citizen to be given access to such files and to check that the facts about him are correct. They have also been trying to limit the movement of personal data from one country to another. Sweden, for instance, took action to regulate the transmission of computerised data across her border as long ago as 1973 (and in 1976, banned the movement of any such information to Britain, such is the scant regard for Britain's laws of privacy).

In the meantime, the collection and compilation of personal information was itself hotting up. The testing ground for the United Kingdom has been Ulster, where the military headquarters at Lisburn, near Belfast, have for at least six years played host to a very powerful computer indeed, which gathers the intelligence collected by military patrols and sifts and stores it. The effect has been that the military forces have for all these years had a very good idea of the movement of men, women, children and cars, have known the company most people keep, their shopping and work habits, and any deviations they might make from what appears to be their normal routine.

The Northern Irish programme has provided the basis of the data retrieval system employed by police forces in England. This in turn has kept a full note of the entries in every constable's notebook; every car registration outside the White Horse the night a felon was known to be inside, everyone the policeman happened to run into on his rounds. Fine and harmless for the man going about his lawful duties, not so good if he is drinking with a lady who isn't his wife.

Privacy is the most emotive issue in the world of telecommunications. Information in the wrong hands can mean power. At its most extreme, the power to blackmail; less extreme, the power to direct a commercial shot which the victim will find it hard to resist because it matches his tastes so closely; in education, it can represent an influence more subtle but nonetheless powerful; in the most commonplace of situations, it can simply mean that the ordinary citizen feels naked, over-seen, uncomfortable.

What the Swedes, and to a lesser extent the Americans, have proved is that strict legislation can remove the worst excesses of the Data Bank Society within countries, but the dawning fear as we move into the eighties is that there is one situation which no legislation in the world can prevent, and that is the international transmission of sensitive data across national boundaries and into countries where there is no legislation at all to prevent traffic in sensitive, and therefore valuable, information.

The 1977 General Assembly of the Intergovernmental Bureau of Informatics put this fear into its own bureaucratic language, calling for: 'Recognition of the transnational dimensions of information and its economic and social implications and consequences to national sovereignty, especially when large amounts of data are transmitted abroad for processing or storage or in cases where foreign interests collect data about a country and its people and do not make it freely available to the government of that country.'

Computer expert Adrian Norman has put it more wittily, if no less frighteningly. Couched in the style of a report by the august company for which he works, international consultants Arthur D Little Inc., he has prepared a fictitious feasibility study for an enterprise called GOLDFISH (Global Online Data Files and Information System Haven). The logic is clear. First, offshore havens have long been available for activities which would be illegal in most countries but against which there are no international regulations: offshore funds, tax havens, flags of convenience, freeports and so forth. No problem, therefore, in evading national laws by setting up an international, offshore data haven.

Secondly, there are no laws banning the transmission of electronic data across national borders. As Norman indicates, the Americans in particular are likely to protest loudly at any proposed legislation which smacks to them of interfering with free trade. Even if there were such a law, it would be unenforceable, because it is so very easy to smuggle data, either coded and hidden among normal transmissions, or in tape or cassette form carried by human envoy. Finally, the market for such data will always exist; it includes not only the credit bureaux already mentioned, but also insurance companies, market research teams, fund raisers, employment agencies, private security companies, newshounds and more besides. The report's conclusion is unequivocal. 'We recommend a full-scale feasibility study. By establishing GOLDFISH soon, they will close the market to the smaller scale operations of illegal data pirates and give themselves plenty of time to buy up large data banks before regulations ban such sales.' It would be scary, if it weren't meant as a joke.

Present Tense

The world's first 'wired city' with telecommunications built into every aspect of its life is being established in Higashi Ikoma, a new city near the industrial centre of Osaka in Japan. Everything we've talked about in this chapter will soon be available to 300 families who will shop by closed-circuit television, pay for their goods by electronic funds transfer (even their gas and electricity and water meters will read automatically through the system and the accounts will be settled without the customer having to lift a finger). They can already summon their favourite television programme from a videotape library of movies and shows, and have it pumped to their own set. The children are educated by computer just like the students of Minnesota, while the telcom system monitors their homes for burglar intrusion or the start of a fire. And one day when the family is in another place, eating out at a restaurant for example, they will be able even to adjust the heating in their own home by a remote control, so that the house is just right when they return.

HI-OVIS, as the scheme is called, incorporates no less than 3000 million yen – £150 million of fibre optics funded and run by the all-powerful MITI, Japan's ministry for international trade. Big business is in there with a vengeance: Fujitsu have installed the computer controls, Sumitomo the fibre transmissions and Matsushita the terminal equipment. The scheme is too new for evaluation; the wired-in citizens/experimental guinea pigs won't complete their 'field trial' till the end of 1979. But what Japan does today, Europe very often does tomorrow, which carries the strong suggestion that the 1980s will bring life by remote control firmly to our own doorsteps. The technology exists, and there are powerful forces driving it in our direction.

Judith Hann

Tomorrow's Babies

Pregnant women today expect to produce healthy, undamaged babies. For this reason, interest has moved on from the safety, to the style, of childbirth.

The 'nature knows best' approach has become popular over the past few years, with more women demanding home confinement and an emotionally rewarding labour. They have turned their backs on maternity hospitals, claiming that childbirth has become too scientific, induced 'convenience' births too common, and that machines are taking over from nature.

But controversy over the best method of delivery and the safest place to have a baby is suddenly intensifying, as a result of the recent Office of Health Economics report which stated that many babies died or were born handicapped because complications during pregnancy were neglected until they became acute.

An increasing number of doctors are now warning *against* the 'nature knows best' approach to childbirth. And parents themselves, particularly those who have experience of babies damaged at birth, are mounting campaigns. The Spastics Society, for example, has collected millions of signatures for a petition urging the Government to reduce the 'holocaust' of baby deaths.

It is being accepted, officially at least, that the safety, and not the style, of childbirth should be the current issue. Britain's childbirth record is poor compared to other countries. In England and Wales, eleven babies in every 1000 are stillborn. And for every 1000 live births, another sixteen die during their first year.

If we could reduce this death rate to levels achieved in Scandinavian countries, where the perinatal mortality rate is only nine per 1000 births, we would save almost 5000 lives a year. And for every life saved, another three babies would be spared from brain damage or some other form of handicap.

We have fallen badly behind other countries. For example, in 1960, the perinatal mortality rate in France was 22 per cent higher than in England and Wales. It is now 10 per cent lower than our rate. Japan, in the same period, has done even better, reducing the perinatal mortality rate of sixty per thousand to only nine per 1000 births.

The most frustrating aspect of this situation is that British technology, aimed at increasing the safety of childbirth, is well advanced. The knowledge and techniques we have, however, are not being applied to all births, because of wide regional differences in availability.

If a woman has her baby at a teaching hospital, stocked with the latest and best in life-support technology, the chance of her baby dying is half the national average and about on a par with the best in the world. But in some areas, including parts of Tyneside, Lancashire and Yorkshire, the risks rise enormously, with a perinatal mortality rate of twenty-five per thousand. In remote rural areas it reaches thirty-five per thousand.

A recent Government report shows that fewer babies would die if money was spent on spreading the existing techniques to British hospitals in all areas. It also calculates that the number of babies leaving hospital handicapped could be reduced by half. One good example of what can be done is in the Edinburgh area, where the cerebral palsy rate has been more than halved, due to combined attacks on prematurity, birth injury, asphyxia and jaundice.

We once led the world in the field of obstetrics, but now lie eleventh in the league table of perinatal deaths. The Secretary of State for Social Services has said that this is nothing like good enough and has asked all area hospital boards to organise better special care services for babies at risk. There is a particular need for intensive-care units. Almost all babies in Scandinavia are delivered in large maternity hospitals with intensive-care facilities, but the few units we have in Britain cannot cope with all the babies requiring specialised medicine. At the moment, they have to turn away at least half of the babies in need.

It has been estimated that 15 per cent of all British babies need special attention, while 2 per cent require sophisticated medicine to survive. Extra equipment would help existing traditional maternity hospitals to look after these babies. Many die through lack of oxygen during labour, and the safest way to check a baby in danger is to use a method of continuous monitoring of the foetal heartbeat through an electrode attached to its scalp. Unfortunately there are times when hospitals have more babies at risk than machines to help them. In some cases, foetal monitoring machines have been bought as a result of local appeals, which is a random method of helping Britain's babies.

One in four perinatal deaths is thought to be due to congenital abnormalities. The number could be reduced by antenatal screening tests, because women who learned that they have an abnormal foetus, could then choose to have an abortion if they so wished. But these diagnostic services are also overstretched.

Our childbirth record is poor because we do not spend enough money on this whole area of medicine. Arguments about cash shortage are weak, because it is shortsighted to limit spending in this field. It has been calculated that although Britain spends only half a million pounds on research into special care for babies at risk and other methods of preventing handicaps, we spend £300 million on caring for all the handicapped people in our population. It is now obvious that we must increase spending on premature babies and others at risk.

Life-Saving

So what techniques are available to reduce the death rate? Firstly, there are several important new methods of dealing with respiratory–distress syndrome (RDS), a problem which causes the death of thousands of babies every year because their lungs are under-developed and therefore do not function properly. RDS occurs when a baby's lungs lack surfactant, a detergent-like substance that reduces tension within the lungs' minute air sacs, helping them to inflate easily and preventing them from collapsing completely after each breath.

The danger with RDS is that the damaged lung cells and fluids from the blood can eventually combine to form a fibrous material, called hyaline membrane, which fills the air sacs and stops oxygen in the lungs reaching the baby's blood-stream.

One technique involves monitoring the baby before birth. The amniotic fluid surrounding the foetus in the womb is examined for a substance called lecithin, which is found in the surfactant. Measurements of this lecithin indicate exactly how well-formed the baby's lungs are.

Doctors are also giving steroids such as betamethasone to mothers, one or two days before delivery. They are believed to prevent RDS, by passing from mother to baby and stimulating the lungs to produce surfactant.

A new breathing aid also helps the baby once it is born. Ordinary respirators are unsuccessful with RDS patients because their lungs collapse with every breath. For this reason doctors are trying continuous, positive airway pressure, a method of giving oxygen under pressure through tubes in the baby's nose or mouth, or even by a plastic hood over its head.

It is vital to monitor the amount of oxygen received by any baby with breathing problems, because too much oxygen may cause blindness and too little can bring about mental retardation. The normal method of checking the oxygen level is for a doctor to remove a small amount of blood from a catheter, passed up through the umbilical artery into the baby's aorta – one of the main blood vessels from the heart.

This blood sample is then put into a machine which registers the oxygen level. The sampling and checking has to be done several times an hour, or even more frequently during the critical first few days of a baby's life. The oxygen given to the baby is then increased or decreased according to the results of each blood test. But the problem with this method is the time lapse between sampling. Critically ill babies often deteriorate within minutes. It became obvious that a method of carefully monitoring the blood continuously, without having to remove samp-les, would help the sick babies and the staff looking after them.

After two years of research, a team at University College Hospital, London, found the solution. The technique, which has now spread to other parts of the

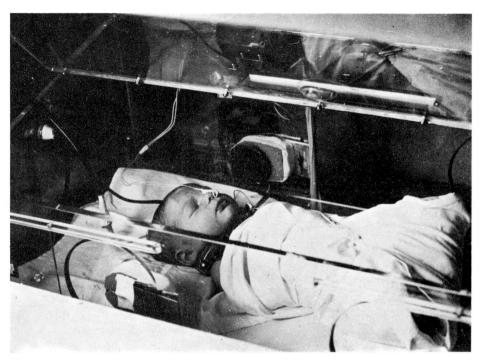

Testing the hearing of very young babies (see page 147) is one of the new techniques which help pediatricians

world, involves a catheter with a silver cap, which acts as a direct current electrode. There are two very thin wires running from a miniature two-pin plug along the catheter to its tip. The catheter electrode is inserted into the baby's aorta through the umbilical artery soon after birth.

The silver cap at the tip is the electrode's anode, connected to one of the wires. Inside the anode, but insulated from it, is the other wire, which acts as the cathode. The whole cap is then smoothed off, coated in electrolytic crystals and dipped into a polystyrene solution, which dries round the cap to form a thin permeable membrane.

It acts like a miniature DC electrode. A small voltage is applied between the anode and cathode, and this voltage induces the flow of current across the top of the catheter. The size of the current flow depends on the oxygen concentration in the blood. Details of the current flow are sent back down the wires to a pen recorder which traces the oxygen level on a chart.

An alarm can be fitted to the system in case this level becomes critical, but most important of all, the continuous monitoring allows doctors to check the condition of every baby at a glance. Oxygen can be adjusted far more quickly, and once the catheter has been set up, the baby's blood does not have to be transfused.

The same medical team at University College Hospital is also developing methods of monitoring oxygen and carbon dioxide levels using electrochemical sensors placed on the skin. This can be done by putting a heated oxygen electrode on the skin surface; either a ring anode surrounding three micro-cathodes or a large cathode which is heated itself. Future non-invasive techniques for monitoring gas levels are expected to involve gas chromatographs and mass spectrometers.

Sophisticated techniques are also important during that dangerous period when a baby at risk is being rushed to an intensive-care unit from its home or a traditional hospital. For that reason, transport incubators have been developed with in-built mechanical ventilators for babies with respiratory problems, and facilities for monitoring body temperature, heart rate and the inspired oxygen concentration.

The Pre-Conception Clinic

There is another, quite different, approach to improving our perinatal mortality figures. Medical Research Council (MRC) doctors in Newcastle have started a new type of service: the pre-conception clinic, where women wanting healthy babies go even before they are pregnant.

The idea is to help women who are bad reproducers. These are the patients who, although they are fertile and conceive easily enough, find that something goes wrong during the weeks of pregnancy. It is a very big problem, because, not only do the babies die at or soon after birth, but thousands more are miscarried every year – in fact almost one quarter of all babies conceived fail to reach term.

Doctors at the MRC Reproduction and Growth Unit in Newcastle believe that many of these babies would thrive if only they were given the right chance. Unfortunately they suffer because their mothers' bodies do not adapt to the extra work of pregnancy. There are patients who are unable to provide for all their babies' needs through the umbilical cord which links mother and foetus, as they develop inside the womb. It is important for mothers to produce a good blood flow through the cord, carrying essential oxygen and food to their babies.

A foetus in the womb is very much like an astronaut out in space. Both have a support system; the foetus is linked to the placenta by the umbilical cord and the astronaut is linked to the capsule by a life-line carrying all his needs.

But there is one difference. Each astronaut has the backup of alternative links and communication with earth, if anything should start to go wrong. The foetus, however, does not have this advantage. As there is no direct communication between it and the mother, it is left to doctors to look for warnings of problems during pregnancy.

Pre-conception clinic doctors will make sure that the life-support system is in full working order before the baby is put into its capsule, the womb. Tests are carried out before a woman becomes pregnant. They are done for three months, during menstruation, because at this time there are small changes which prepare the body for the possibility of pregnancy.

The changes measured include the increase in the amount of blood flowing around the body, extra urine produced by kidneys which are working harder, and chemical changes, including less salt in the blood. In general, the body does more work during menstruation, which acts like a springboard to possible pregnancy.

The tests are simple. Blood volume is measured using a blue dye. It is injected into one arm in measured amounts, and once in the bloodstream it attaches itself to protein in the blood. After it has been in circulation for about ten minutes, a blood sample is taken from the other arm. The more diluted the colour, the greater the patient's blood volume.

This and other simple tests on urine and blood samples, done on days 7, 14 and 21 of each menstrual cycle for three months, give details of hormone and chemical changes in the body. The women who do not make these changes, who cannot adapt their bodies for pregnancy, are the bad reproducers. MRC doctors hope to help them using doses of hormones, already available, to adapt their bodies for healthy pregnancy.

The Baby at Home

When a mother and her new baby leave hospital it is still important to check progress. Professor John Emery, of Sheffield Children's Hospital, has developed a 'risk-rating' scheme, which has helped to bring the local infant mortality rate down to Scandinavian levels. Babies who are possibly at risk are monitored by health visitors, who spend time with mother and baby, watching for danger signs. It seems significant that cot deaths are often preceded by mothers' worries, which they do not tell anybody about.

Lucky parents may find a GP who checks new babies in his practice. One of the rare doctors to do this, Graham Curtis Jenkins, was filmed at work in his busy, semi-industrial practice in the south-east of England by *Tomorrow's World*. He sees a baby, preferably with both parents, after birth, at seven months, and then at yearly intervals until the child is five. His methods of surveillance, screening and assessment produce excellent results. He can diagnose problems and difficulties at an early stage, usually long before they become evident to parents. Specialist help is then brought in, before handicaps develop and harm the development of the child.

But a child's very first visit to Dr Curtis Jenkins is probably the most important. He spends thirty minutes with every parent and baby as soon as possible after they leave the maternity hospital, and sets about proving that newborn babies are very able. He illustrates their ability to see, hear and respond, often to the amazement of parents. At least 80 per cent believe that their babies can neither see nor hear, that after birth they are at first like 'blobs – witless tadpoles'.

He soon proves them wrong by showing that they can recognise faces, hear clearly and behave in a sensitive manner. He tells parents to talk, sing and respond to their babies, and explains that this is the best way to build up close relationships right from the start.

Work of this kind will reduce the number of parents who stunt their babies' development because they underestimate their abilities at birth and in the first months of life.

'Every child is born a genius,' said the prophetic American, Buckminster Fuller, to emphasise his belief that few of them ever reach their potential because they are underestimated by the adult world. Parents can restrict early development by treating babies rather like padded alimentary canals, concentrating on feeding and cleaning them, while putting too little emphasis on their emotional and mental needs.

The common attitude is that good babies are quiet babies, so they are usually put away to rest in their own rooms. But now certain doctors are telling parents that babies need less sleep than most adults believe, and that they are happier and more stimulated being in the main room where everything is going on. The idea of leaving babies to lie on their backs is also changing. Babies in relaxer chairs, where they can see more and move their limbs, are thought to develop faster.

So babies are more capable in the early weeks than parents normally believe. It is also thought that they learn better at the start of life than at any later stage in their development. This ability is best made use of when there is a close relationship between the baby and an adult. This is normally the mother in our society, although there is no magic bond. Any other caring, stable adult will do just as well.

It is important to foster the relationship from the start of life, and research has shown that demonstrative mothers form stronger and more immediate attachments to their babies. In hospital, and later at home, these demonstrative mothers have the most active babies. It is like the problem of the chicken and egg. The more active babies produce stronger maternal feelings, while maternal, attentive mothers encourage their babies to respond. The quality of this relationship is vital because it is this relationship which assists the emotional and intellectual development of the child.

It is now thought that the methods used in some modern maternity hospitals may restrict the growth of this relationship between mother and child. For so-called medical reasons it is common to separate babies from their mothers after birth for up to three days. In the case of premature babies and problem births, it is done to monitor a child's early development. In other cases, it is to give women a rest after difficult births.

Dr John Kennell and Dr Marshall Klaus, at Case Western Reserve University, Cleveland, Ohio, decided to find out if early separation from the mother caused the same sort of problems scientists have seen in animals. With goats, for example, mother and infant form a close bond in the first five minutes of a kid's life. If removed before this time, nearly half are later butted or kicked to death by their mothers.

Kennell and Klaus studied two groups of patients for two years. Mothers in the first group were separated from their babies for the usual three days, apart from feeding. The mothers in the second group, however, were given their babies in bed for an hour after birth, and allowed to have them for five hours each day, on top of feeding periods, during those first three days in hospital.

Differences in behaviour were striking, even after only one month. Mothers allowed close contact with their babies from the beginning spent three times as long as other mothers cuddling and kissing them. They also held them closer while feeding, and were more gentle and soothing than the other group of mothers.

Despite this important research, some maternity hospitals and nursing homes still have central nurseries where babies are taken after feeding. Fortunately many are changing this policy, and allowing babies to stay in their cots by their mothers' beds. But, at Charing Cross Hospital in London, one consultant, Dr Hugh Jolly, has gone much further. He now believes in letting his patients have their babies all the time, not in cots, but actually in bed with them.

It is interesting that a few years ago this paediatrician strongly warned parents against taking children into bed with them. Now he has reversed his point of view and encourages mothers on his wards to sleep with their babies.

He has changed his mind because he believes it helps to strengthen the bond between mother and child, and because many patients have always wanted it this way. He also advises his patients to carry on with the habit at home; to have what is called a family bed, which provides contact, warmth, love and security.

Families who have never tried sleeping together often think it would be impossible to get a good night's rest. But fans of the family bed idea say that physical contact makes babies far less troublesome at night. They say it can have a long-term effect in that unnecessary fears about being alone and being in the dark may never occur.

One of the main objections to the family bed is the danger of 'overlying'. Dr Jolly claims that he knows of no scientific evidence to point to overlying as a cause of infant deaths, and he does find that parents who have suffered a previous cot death in the family often want to sleep with their babies.

We showed the results of time lapse photography on *Tomorrow's World*, which does illustrate how the parents move and alter their positions during sleep to avoid overlying.

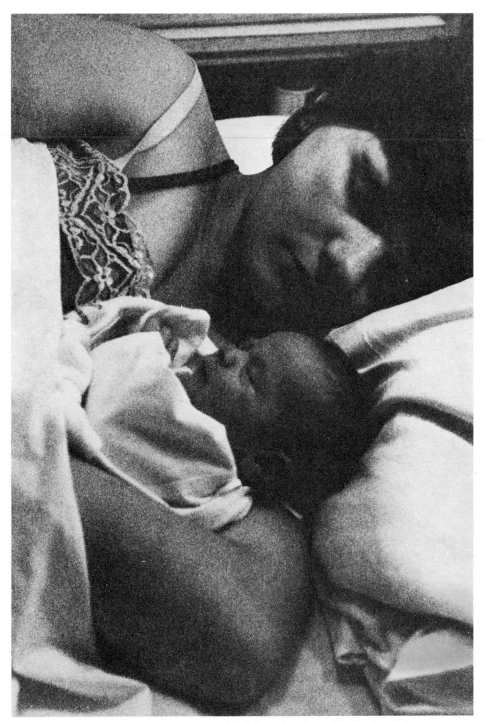

Newborn babies are now encouraged to sleep with their mothers in one London hospital

There are two other possible problems – a lack of privacy for parents and the difficulty of weaning a child away from the family bed when the right time comes. But the family bed is growing in popularity as our attitudes to children change. More parents are rearing them with love and sympathy, instead of with a schedule in one hand and a clock in the other.

Child-rearing habits go in cycles. A century ago the family bed was advised. Dr Pye Henry Chavasse, the 'Dr Spock' of his day, said: 'A baby needs the warmth of another person's body.' Since then so much has changed that parents often feel guilty when they take their child into bed with them. Dr Jolly is suggesting that there should be no rules.

Is Every Child Born a Genius?

Far more research is now being done on the behaviour and abilities of babies and young children. And the results are confirming the beliefs of doctors like Curtis Jenkins and prophets of the Buckminster Fuller persuasion. Even newborn babies are very able. They are not just a bundle of reflexes, as many scientists and parents believed in the past.

It has taken a long time to prove because babies are very difficult to test. Although sophisticated things may be going on inside the babies' heads, their control over their bodies is so primitive, that it is difficult to interpret their responses.

The work at Bedford College, London, which showed that babies learn to recognise their mothers' voices within the first four weeks of life, illustrates the problems of research of this kind. The researchers had to exploit one of the few motor skills that is well within the smallest baby's repertoire, that of sucking.

The method used was to let the baby hear its mother's voice every time it sucked a dummy. Babies will suck dummies even when they are not hungry, so apparatus was arranged with a dummy connected to a light switch. Every time the dummy was sucked it switched on a light, that was invisible to the baby, but could be seen by a woman sitting behind a screen. When she saw the light come on, she knew it was her cue to begin reading.

It did not take long for the babies being tested to learn that sucking led to being read to. Once they had grasped this, they were tested on the relative amount of time they were prepared to spend sucking in order to hear either their mother's voice, or the voice of a stranger.

The research indicates that babies have a definite preference for hearing their mother's voice. It is even more interesting, however, that babies less than one month old could tell the difference between voices behind a screen.

The ability to imitate facial gestures has been considered a landmark in a child's development for many years. Jean Piaget, doyen of child psychologists, said that children reached this stage at eight to twelve months, and that before

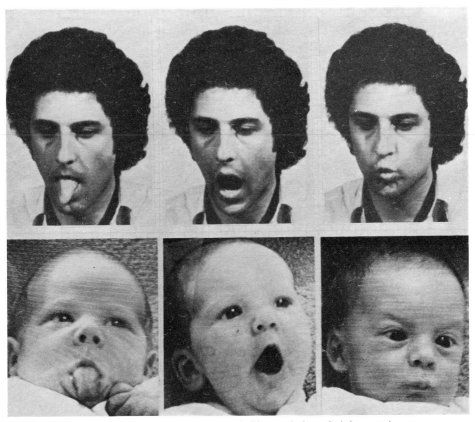

American research has shown that even very young babies can imitate facial expressions

that age they lacked the perceptual and cognitive skills to match a facial expression they see with one they make of their own.

He has now, however, been proved wrong. American scientists have shown that even babies only twelve days old can imitate facial gestures very accurately. Child psychologists are now reconsidering their theories of cognitive development in the light of these findings.

They also proved the many parents wrong who believed their newborn babies could not see anything more than blurred shapes. Research at Cambridge University is also confirming that vision is well-developed at birth. When babies being tested are offered a choice of fuzzy or clear pictures of faces, they prefer to look at the pictures offering as much detail as possible. Babies up to five or six weeks old can tell one face from another by outlines; the shape of the face, a beard or the hair. That is why a baby may become alarmed if his mother suddenly changes her hairstyle or wears a bathcap. And it explains why very young babies spend a lot of time scanning the outline of faces.

At seven weeks and older, however, babies prefer eyes. A recent report by three American psychologists, shows that babies at this age spent nearly 90 per cent of a test period looking at adults' faces, particularly at the eyes.

The research suggested that the babies were not attracted to the eyes merely because of their physical properties, like colour, movement and contrast. If the babies had been interested in the faces' physical attributes, they should have spent more time looking at the mouth when the adults were talking to them. The fact that the eyes got most of the attention, led the psychologists to suggest that by intensively scanning the mother's eyes, especially when she is talking, the baby may be encouraging her to continue to speak and pay attention to him or her.

Photorefraction

Work at Cambridge has also solved that difficult question of whether babies can focus their eyes at different distances. A device called the Howland Photorefractor provided the answer. It fixes on to the wide-aperture lens of a 35 mm camera and contains a fibre-optic light guide which leads the light of a flashgun to emerge in the centre of the lens. Surrounding the fibre-optic are four segments of a cylindrical lens, arranged like the vanes of a windmill.

If the eyes are in focus on the camera, the flash makes an image at the back of the eyes and the light from this bounces back along the path it came. Very little light spills over the lens segments to get back to the film. But if the eyes are out of focus for looking at the camera, the image in the eye is blurred and a large patch of light is reflected back to the camera lens.

a *A young baby's eyes being tested using the Photorefraction method*
b *The special camera attachment has a miniature flash in the centre of the four-part lense*

a

b

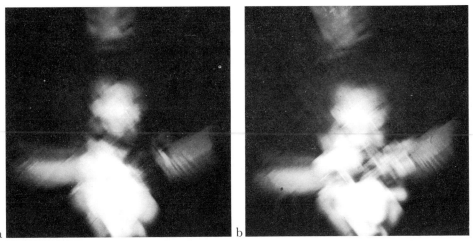

a   b

a   *The light reflected back from the baby's eyes is formed into a star-shaped image on the camera file. If the arms are short, it means the eyes are in focus*

b   *If the arms of the star are long, the eyes are not in focus*

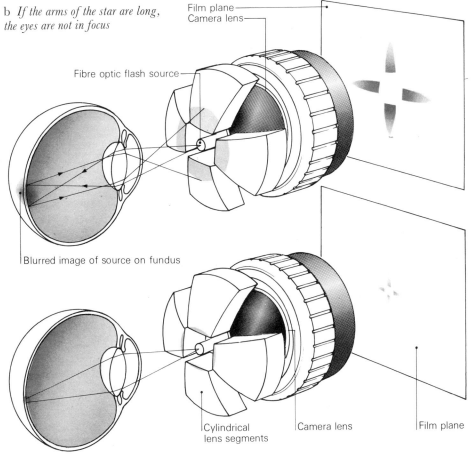

Because of the way the lens segments are cut, the light falling on each segment is collected into a streak, producing a star-like image of each eye on the film. The length of these star arms can be measured. Short arms means the eye was focused on the camera, long arms mean it was focused at too near or too far a distance.

Testing vision with the photorefractor is simple. A baby sits in front of the camera; someone attracts the baby's attention to a brightly coloured rattle by the camera for a moment, to set off the flash. The picture can then be repeated with the rattle at different distances. The flash is quite safe to look at, because it has as short a duration as an ordinary snapshot flash, and is also much fainter.

The photorefractor can prove whether there is a sharply focused image in the baby's eyes. But that does not necessarily mean that the baby can see a sharp picture. The image in the eye has to be signalled to the baby's brain, and the baby's nervous system is not yet mature. So the connections that carry the message from the eye to the brain may not be complete enough for the baby to see all the detail of a sharp image.

But the Cambridge researchers check what the baby sees by testing what he will react to. The baby can be shown sharp and blurred pictures side by side, and fine stripes next to a blank screen. If he consistently prefers to look at the sharp picture and the fine stripes, it shows that he is not just seeing a blank blur.

The combination of tests has shown that babies can pick out the main features of a face from between birth to one month old, but that between one and three months old their vision becomes much sharper. Photorefraction shows that babies under two months can focus over as wide a range of distances as adults.

Most babies show astigmatism, which means their eyes cannot bring both horizontal and vertical lines into sharp focus at the same time. This is not nearly so common in adults, which means it must clear up naturally, although it is not yet known when.

The photorefractor has shown scientists a lot about the normal development of vision in babies and indicates that even newborn babies are very able. But it is also being used to investigate babies who have problems, like squints. The great advantage of photorefraction is that it is a non-invasive technique. Before it had been invented, a baby's focusing power could only be checked by looking into the eyes with a retinascope. This involves using eye drops to relax the focusing muscles thereby preventing any changes of focus during testing.

The new method makes it much easier to measure focus without using drugs. It makes it easier to detect any defects in babies' sight at an early stage, when it is simpler to correct them. And it is also useful in telling doctors whether or not treatment is effective.

The Listening Muscle

It is also important to check a child's hearing at an early age, because if normal speech is not heard in the first few years, a child will never learn to speak properly, even with the help of hearing-aids and speech therapy. The problem is that although one in 1000 children are born deaf, most are not discovered until they are three – when it may already be too late to correct their speech inadequacies.

The reason for this is that it is very difficult to check whether a baby or young child has perfect hearing. As I have already explained, it is hard to interpret a baby's responses because he has such poor control over his body at an early age.

The routine hearing tests using rattles, cup and spoon and other noises, are very much a hit and miss affair. Many children fail to respond to the tests, not because of faulty hearing but because they are ill, distracted by something else, or just plain bored.

There are other more sophisticated tests. One involves putting electrodes on the skull. A noise is made and the electrodes measure any electrical impulses in the brain. But at any one time there are many varied signals shooting around the brain. Most of them are nothing to do with hearing. The child being tested has to sit still, with the electrodes in place, for up to two hours while the signals are sorted out. Hardly the ideal hearing test.

A more recent method, called cochleaography, has to be done under anaesthetic, because a needle is pushed through the ear-drum until it touches the cochlea in the middle ear. A noise is made during the test, and doctors then check to see how much of the sound is picked up in the cochlea. It is accurate, but far too complicated to become a standard test.

Finding a simpler method of checking hearing means going back into man's primaeval past. If you watch the way animals behave when they listen for a noise, you will realise that most of them prick up their ears at the slightest unusual sound.

Doctors at Guy's Hospital, London, decided to test the theory that man once had the ability to prick up his ears. They studied a muscle just behind each ear which apparently serves no useful purpose today. They found that a signal passes from the brain to the muscle every time there is a sound. That signal is almost certainly an instruction to move, although nowadays the muscle is far too weak.

But, the fact that the brain sends the signal has been used as the basis of a new hearing test. Simple electrodes are stuck behind each ear on the now redundant muscle. They detect the slightest signals coming from the brain. A noise which sounds like a series of clicks is made near the child being tested. It does not matter if the child is sleepy or distracted; if he hears the clicks the brain will send a signal which will be picked up by the electrodes and monitored on an oscilloscope.

The test starts with loud clicks which become progressively fainter. A pattern on the oscilloscope gives doctors an accurate reading of each child's hearing threshold. No anaesthetics are needed for the test, and it takes only a few minutes. But perhaps the most important point is that there is no lower age limit. It would be quite practical to test all newly-born babies before they leave hospital. This, combined with modern hearing aids, could mean a massive reduction in speech problems through deafness.

Choosing your child

Future research will help the people who desperately want to choose the sex of their children. Work on this intriguing subject is still at an early stage, although theories abound. One is that the food you eat can determine the sex of your child. The belief that mothers who eat cheese, eggs, pasta and fish, all without salt, give birth to girls, is actually being tested. Women who like salty food, such as ham and Marmite, are thought to have boys.

The science behind this odd theory is that X and Y sperms respond differently to the ratios of sodium, potassium and calcium in the mother's body. Male sperms, which are Y, react well to a salty substance. Volunteers who want girls are being told to reduce their sodium and potassium intake, to take less salt, but to increase their calcium intake by drinking more milk.

Another similar idea – that X and Y sperms respond differently to the chemical environment of the vagina – has led to one British doctor patenting a sex-choice spray following some experiments with rabbits. He has developed a jelly-like substance which when sprayed on the females, reduces the acidity of their sexual organs and increases the chances of their offspring being male. This work is based on the observation that sperms carrying the male Y chromosome swim more slowly in an acidic solution than sperms carrying the X chromosomes. The patent proposes a kit of two different gels, one buffered to a pH of 7 and the other to a pH of 7.6. The gel is introduced into the vagina using a special tube just before intercourse – an acid gel if you want a girl and the alkaline gel for a boy.

The acid/alkali content of the vagina alters naturally at different times of the month, which could mean that X and Y sperms have a better chance of survival at certain times. One leading British expert believes this, and tells patients to time intercourse usefully – early in the month for a boy and late for a girl. He also claims that the more frequently you have intercourse, the more likely you are to conceive early in the cycle, and so have a boy.

The repercussions could be considerable if any of these methods were found to work. Having a son is preferable in Third World cultures, where parents go on having children to ensure survival of a male heir. Being able to choose to have a son might cut the size of the family and reduce the birth rate. One must hope that there will not be more sinister results when man controls this delicate balance.

William Woollard

Outer Space

Space is no longer the new frontier. Gone are the days of the great television spectaculars, following every detail of the launch; men bouncing around on the surface of the moon, and the astronauts' return, coming back almost like messengers of the Gods returning from Mount Olympus bearing their cryptic messages. Phrases that once carried with them a thrill of excitement – 'mission control', 'count down' and 'lift off' already seem dated; part of yesterday's language despite their continued use. NASA has become just another American agency striving to justify the continuation of its programmes.

The great Apollo circus is over, brought to an end not so much because the scientists believed they had achieved all their objectives, but because the American public began to lose interest at about the same time as they became aware of the colossal size of the bill. Bob Hope expressed it more wittily than most perhaps when he said that if they had taken the cost of getting a man on the moon and piled the notes one on top of the other, 'they could have stood on the goddarned dollar bills and *picked* the rocks off the moon'.

So what's left, apart from a pile of grey moon rock big enough to keep universities and institutes all over the world busy well into the 1980s? Well, in place of showbiz-style entertainment there is now work; exploitation instead of immensely courageous exploration; scientists and engineers working out in space instead of astronauts. And they are opening up all the possibilities of mineral mines in the mountains of the moon and solar power stations endlessly circling the earth. That is indeed what is happening in near space. But in parallel with that near space exploitation there is also the great quest to explore deep space. Strange, angular, silver-wrapped probes packed with sensors and monitors and recorded human voices in place of men are sent hurtling silently out towards the edge of the solar system; out towards Jupiter and Saturn, and beyond into the blackness of space, while their computers bleep a continuous stream of data back to earth to tell us what they have seen. Out of the experience of the lunar orbit, for example, has come the realisation of the 'Grand Tour'.

Deep Space Exploration

The Grand Tour has been a dream that has been nurtured by space scientists for decades. Essentially it involves a long looping journey through the solar system, using the gravitational pull of one planet to accelerate the space probe and hurl it, like a stone out of a slingshot, on to the next one.

In the summer of 1977 two Mariner Class spacecraft called *Voyager I* and *II* were launched on a fifteen-year journey that carried them out to the giant planets of Jupiter and Saturn and beyond to Uranus. It is a project that is designed to fill some of the giant gaps in our knowledge of these darker, colder, outer margins of the solar system and – hopefully – to provide one or two more clues to the earth's origin and its distant history.

Although the spacecraft were carried aloft by the immensely powerful Titan-Centaur rocket, they were the first planetary spacecraft to carry their own solid fuel propulsion motors. Curiously, it was *Voyager II* that was launched first. But because *Voyager I* flies a faster trajectory, it soon made up the twelve days' difference in launch date and overtook *Voyager II* on 15 December 1977.

But the completion of the Grand Tour depends entirely on what the space engineers call 'gravity assist'. When the space probes approached Jupiter, the planet's gravitational field first accelerated the spacecraft around the planet, and then hurled them off towards Saturn. For the mission to succeed both planets have to be in a particular alignment, an alignment that they only take up once every forty-five years. The next Grand Tour will be planned by the grandchildren or even the great-grandchildren of today's NASA scientists.

Because of this slingshot effect, *Voyager I* will arrive at Saturn in the late summer of 1980, about three years after launch. Without it the 2200 million kilometre flight from earth would have taken over six years. One of the main problems has been power. Because the spacecraft are heading directly away from the sun, solar panels, the normal source of power for spacecraft, would soon become too feeble to power the equipment on board. So they are carrying their own power supply – nuclear power.

Three radioisotope thermo-electric generators (RTGs) are mounted on a boom at the base of each spacecraft. They convert the heat released by the radioactive decay of Plutonium 238 into electricity, and will be able to supply about 400 watts of power continuously throughout the life of the mission.

They reached full power only eight hours after the launch. During pre-launch operations and until about one minute after lift off, the generator interiors were kept filled with an inert gas to prevent oxidation of the hot components. A pressure-relief device emptied the generators of gas as the spacecraft passed through an altitude of about 6080 metres on their way out of the atmosphere. Twenty minutes later when the primary and secondary stages had separated and the spacecraft were settled down on course the RTGs took over from the batteries and began to supply power to the instruments.

All went pretty much according to plan until about two months out, when the *Voyagers* were in what is called the 'engineering cruise phase'. This is the stage of the mission when the engineers back at control run through an exhaustive series of

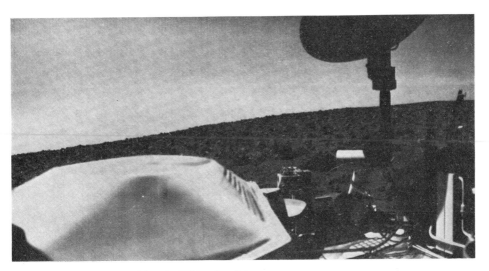

This picture was taken on Mars by Viking Lander 2

Now known not to be the only planet to have rings, Saturn still remains the most spectacular – and is big enough to swallow up 200 earths

checks and tests – to set the spacecraft up for their planetary encounters. It soon became obvious that they had a major snag with *Voyager I*. They were unable to achieve full deployment of the package of instrumentation. Immediately they went into a series of routines to pinpoint the source of the problem; interrogating the spacecraft itself, as well as adopting a technique used to great effect during the Apollo missions – working on a *Voyager* twin at the Jet Propulsion Laboratories (JPL) to see how they might get around the disability.

In 1978 the engineers were still working on freeing the scanning platform because they were concerned about the greater quantities of fuel they will use up in having to orient the entire spacecraft at the planets, instead of just the platform and its instruments. While the controllers were wrestling with *Voyager I* an unusual thing happened. They overlooked *Voyager II*. They omitted to make contact with the second spacecraft for the routine weekly check. This sent *Voyager II* into a minor electronic trauma. Not receiving any signals from command at JPL it assumed that something had gone wrong back at base – so it switched to its on-board back-up programmes in order to continue with its mission. When mission control in Pasadena eventually got round to contacting *Voyager II* it received a very cool reception. It was as if the spacecraft had to be persuaded that mission control was precisely what it was claiming to be. *Voyager II* sometimes obeyed mission control, sometimes its own on-board computer. And it is rumoured that on at least one occasion when JPL asked *Voyager I* to do a 360-degree scan to check some of its instrumentation, *Voyager II* sent back the curt response 'Why?'

There cannot have been many occasions in the brief history of space exploration that mission controllers have simply forgotten about one of their highly cherished, multi-million-dollar packages of instruments hurtling through the solar system. However, these operational problems were overcome and *Voyager I* has completed a spectacularly successful journey.

Until recently our knowledge of Jupiter and Saturn was only rudimentary despite the fact that these two planets dominate the solar system. The diameter of Jupiter is ten times that of the earth and it contains more matter than all the other planets put together – in fact over 90 per cent of all the matter in the solar system excluding the sun. Jupiter's massive gravitational attraction exerts its influence on every other planet and wrenches comets from their orbits. With its retinue of thirteen moons, four of them as large as the smaller planets, Jupiter is virtually the centre of its own miniature solar system. In fact Soviet scientists have worked out that Jupiter will become a sun three billion years from now. Already it is radiating more energy than it receives from the sun.

Saturn also has a family of satellites and it has two quite unique features that have long exercised a fascination over astronomers.

One is a spectacular ring system which appears to be composed mainly of ice

Voyager heads for Uranus – a thousand million miles beyond Saturn

crystals and rocks which measure up to a metre across. The second is a large satellite, only slightly smaller than Mars, that is known to have an atmosphere of methane that may be as dense as the earth's atmosphere.

In spite of their vast size, Jupiter and Saturn are both enigmatic planets, giant balls of gas with no apparent solid surface, composed largely of hydrogen and helium; they represent the original materials from which the sun and all the planets were formed. For centuries astronomers, physicists, scientists of many disciplines, have longed to examine them more closely. It has taken 200-odd years of industrialisation for one nation, America, to develop the wealth and the technological power, to enable them to do so.

Voyager I has now threaded its way neatly through the five inner satellites: Callisto, Ganymede, Europa, Io and tiny red Amalthea, and has streaked past Jupiter itself in a great curving arc. The cameras on board have functioned impeccably, sending back a dazzling stream of photographs; of volcanoes caught in the moment of eruption on Io, of strange rhythmically-spaced ridges on Callisto, and of clouds swirling away from the Giant Red Spot.

The entire visible surface of Jupiter is made up of multiple layers of clouds, composed mainly of ammonia ice crystals coloured by small amounts of materials of unknown composition, and the Red Spot is now seen to be one of the coldest places on the planet and soaring perhaps 25 kilometres above the surrounding layers.

It is already clear that Jupiter, like Saturn, is a ringed planet; *Voyager I* has picked up a thin, flat ring of particles perhaps 30 kilometres thick; but the wealth of information that has been sent back on this enigmatic planet will keep analysts and astronomers busy for many years.

End of story? Not quite. In the early 1990s the *Voyager*s will cross the outer boundary of the solar system, although by then they will have stopped tracking scientific targets and indeed may well have run out of fuel. As they hurtle out into the darkness beyond the system of which we are a part, what are they likely to encounter? Well the short answer is – not a great deal. It will be 40,000 years before they pass the next star at a distance of about one light year ($9\frac{1}{2}$ trillion kilometres). After that it will be 147,000 and 525,000 years before they meet up with anything of further interest.

But just in case this twentieth-century castaway bottle should one day be washed up on the shores of some distant solar system, and be retrieved, NASA scientists have put a message inside. It includes 116 digitally encoded photographs, which apart from the more obvious earthly images, include snaps of the Taj Mahal, a child being born, a snowflake, and a tree toad. It also carries greetings in dozens of languages including Gujarati, Telugu, and Nguni. There are recordings of earth sounds including whales, train whistles, chimpanzees and that uniquely human sound – laughter. Finally on this longest ever playing record of 'Mankind's Greatest Hits' there is a sequence of musical pieces representing the work of Bach, Beethoven, Stravinsky and Chuck Berry.

The Nomads Return

In rather the same way that the *Voyager* mission was able to capitalise on a timely alignment of the planets, the return of Halley's Comet in 1986 could provide an opportunity for astronomers to get a close-up view of what a comet actually is, using a bizarre form of space machine.

Halley's Comet has been observed crossing the heavens since 467 BC which was the first sighting to go down on the human record sheet. It is named after Edmund Halley, Astronomer Royal at Greenwich. On 15 August 1682, when nobody understood that comets were part of the solar system, a comet appeared in the sky. A close friend of Halley's, Isaac Newton, was in the process of formulating some startling new theories on gravitation. Using Newton's laws Halley began to wonder whether the three bright comets observed in 1531, 1607, and 1682, were one and the same. In 1705 he delivered a paper in Latin to the Royal Society in which he went out on a limb and suggested that this same comet would grace the heavens in 1758. Sadly Halley died in 1742, but on Christmas night 1758 an amateur astronomer in Germany sighted the comet, and Halley's name was immortalised.

Of all the comets in the sky
There's none like Comet Halley,
We see it with the naked eye,
And periodically.

The difficulties surrounding any proposal to send a spacecraft to rendezvous with Halley's Comet are formidable. The main difficulty is that it is going the wrong way.

If you imagine yourself looking down at them from above, the sun and all the planets in the solar system rotate anti-clockwise. Halley's Comet, however, has a clockwise orbit. What that means is that to try to get a spacecraft close to the comet would need an enormously powerful and therefore impossibly big and expensive rocket. To catch up with the comet as it zipped through the solar system at speeds of up to 198,000 kilometres per hour would take prohibitive amounts of fuel. Hence the design of a radically new type of spaceship. It will be powered by solar electric rockets, or Ion Drive motors. Panels of solar cells 150 metres long would be extended either side of the craft to generate more than 100 kilowatts of electric power. This would be used to ionise mercury vapour in the main chamber of the ion engine. The ions would then be focused by electric and magnetic fields through a pair of electrode grids, to be expelled in a steady flameless violet exhaust. The exhaust would provide the thrust. The fuel efficiency of this sort of rocket is ten times that of conventional technology. The problem is that the thrust produced is puny, less than a twentieth of a kilogram for each engine in the cluster. So the idea is that for three and a half years after its launch from the space shuttle in March 1982, the Ion Drive spacecraft will race round the sun, steadily picking up speed and gradually changing the angle of its orbit. By late 1985, when Halley's Comet will cross the orbit of Mars, travelling in a clockwise direction, the Ion Drive spacecraft will also be travelling clockwise round the sun. Then during the first half of 1986, space scientists would hope to be able to manoeuvre this curious-looking craft into formation with the head of the comet – some are saying they would be as close as two kilometres. You can imagine the spectacular close-up pictures we would get of one of the solar system's greatest personalities.

Take me to your leader

Of all the space projects manned and unmanned that have taken place since that first balloon-sized sputnik was lobbed into orbit, perhaps the most significant was the *Viking* project. True, it may not have had the televisual glamour of men actually walking and working on the surface of the moon, but in its own way it too involved giant steps for mankind.

The technical task of controlling a fragile craft across 540 million kilometres of space and landing it gently on a prearranged site on a different planet is almost too complex to be contemplated. At *Viking* Control it was the enormous success of two successful landings at different locations on Mars that produced an extra-ordinary emotional 'high'.

After that, carrying out the actual experiments for which the landings were made was almost a flat period, an anticlimax, however significant its results were likely to be to the scientists. Perhaps more important were the questions that the *Viking* Lander asked: routine questions, such as age, geological composition, climate, and chemical make-up. And of course the not so routine question, 'Is there life on Mars?'

Rumour has it that fifty years or so ago a celebrated newspaper publisher sent a telegram to a famous astronomer. 'Wire collect immediately, 500 words on whether there is life on Mars.' The astronomer dutifully replied 250 times: 'Nobody knows, nobody knows, nobody knows. . . .' The same telegram sent today, two years after the Viking mission had carried out all its planned experiments, would elicit from an honest astronomer very much the same response. The debate goes on as scientists continue to work away at the data that the *Viking* craft sent back.

The experiments which were performed yielded enigmatic results, results that indicate the *possibility* of life but not conclusively so. But bearing in mind that *Viking* had only four potential landing sites and two actual locations in which to conduct its survey, it can in no way be said to have comprehensively answered the age-old question. Dr Carl Sagan of Cornell University, for example, has suggested that there might be life on Mars in places where the rather kinder conditions that scientists believe prevailed on Mars long ago, and which may still exist.

A second suggestion is that Martian microbes may have developed hard outer shells to protect themselves from the rigours of the environment; the enormous amounts of ultraviolet light, for example, and the absence of water. So they may have been there all along, it is suggested, during *Viking*'s biological experiments, but protected from detection by that inscrutable shell.

We shall have to wait until mobile laboratories can wander at will across the Martian landscape, sampling as they go – and such a mission has been proposed, called Mars 84. In 1977, when we visited the JPL in Pasadena, they were developing the computer hardware for the vehicle, and not only could it move around and scan its environment, but it could also think. It could, for example, select the potentially interesting from the ordinary rock sample, remember obstacles and learn to avoid hazards like cliffs and trenches.

The point is that with Mars 540 million kilometres away, it takes twenty minutes for a command signal to reach a robot vehicle, and twenty minutes for the signalled response to get back to earth. With information transmission working at that rate it would take hours to complete a single co-ordinated activity such as selecting and picking up a rock. So for such a mission to be worthwhile, the Mars Rover as it is called, has got to be a thinking intelligent beast. It might get a single command 'Move 5 metres North, and pick up the small green rock'. From then on the vehicle would be on its own, using its camera-eyes and its laser ranging devices to work out how to move safely from where it is, to where it needs to be.

So when it eventually lands on Mars and starts meandering pensively across the terrain we shall probably still be unable to say whether or not there is life there. But we shall with some justification be able to say that there is intelligence there.

Project Ozma

In the late 1950s a young radio astronomer working at the Green Bank Observatory in West Virginia calculated the now famous Green Bank Formula. It is a long and complicated piece of statistical processing with rather too many 'ifs' and 'maybes' for it to be used with any certainty. But it did throw up an intriguing and provocative result. Out of the ten billion or so sun-like stars in our galaxy alone, there might be as many as a hundred thousand civilisations which are actively trying to communicate with us.

That young astronomer, Frank Drake, is now director of the world's largest radio telescope. Twenty years ago he pioneered Project Ozma, named after the land of Oz in *The Wizard of Oz* stories. They listened out at the wavelength associated with the hydrogen atom, 21 cm. The hydrogen atom is the most abundant element in the universe and the assumption was that other civilisations would recognise the significance of this 21 cm wavelength and would themselves be transmitting and listening at this point on the radio spectrum.

It has been described as being as effective as wandering over to a haystack, idly picking out a handful of hay, looking through and hoping to discover a needle. Of course if you don't find one you're still no further forward in answering the question 'Is there a needle in this haystack?' That was *Viking*'s problem; it was also project Ozma's. They found nothing. Since then there have been about eight searches, most of them in the last few years, using powerful equipment such as the National Radio Astronomy Observatory telescope at Arecibo in Puerto Rico which is able to monitor three thousand frequencies at once.

But despite the conspicuous lack of even a reverse-charge call from Tau Ceti or Epsilon Eridani, the search goes on. NASA is now planning a systematic and detailed scanning of the entire galaxy that may go on for a hundred years. But even a century-long radio watch on the heavens is almost mundane when compared with a project that has been dreamed up by a team of British scientists.

It is codenamed Project Daedalus, and what they propose is to send a giant spacecraft winging all the way out across space to Barnard's Star to try and discover whether there are planets in the system, and if as they suspect, there are, what sorts of biology have developed there. It will be a journey across six light years. It can only be attempted by constructing a radically new form of spacecraft, out in space. It would use components sent up from the earth and from manufacturing sites located on the moon. It is assumed that they will exist by the middle of the twenty-first century, which is when this project is scheduled for lift off. Daedalus would use fusion power to drive it at speeds up to 12 per cent of the speed of light – 36,000 kilometres a second, with 250 fusion explosions per second being ignited in the enormous reaction chamber. Even so, the journey would take something like fifty years. The scientists who launched it as young men would read of the approach to its destination as they rocked peacefully on their retirement verandahs. But then of course very much the same thing is true of the search for fusion power here on earth – it will last longer than a man's working life; the modern equivalent perhaps of building a medieval cathedral. As this vast spaceship approaches Barnard's Star it will scan the area for planets. Streaking through the Barnard System separate probes will be despatched from the mother ship to take a closer look at anything of interest. By this time, as a result of the nuclear explosions and effect of gravitational acceleration, this armada of terrestrial hardware will buzz the alien system at 13 per cent speed of light. If there is anybody there taking a quiet evening stroll we will have to hope they do not have the facial mechanisms for blinking because if they do they will miss us. Daedalus will pass by like a flash of light on the retina.

But then they may not have faces at all. Scientists on earth are at least agreed on one aspect of contact with extra-terrestrial beings – there is simply no way of predicting what sorts of life forms would evolve on planets unimaginably different from our own.

Project Daedalus may seem like way-out science fiction, but its designers are sober-minded scientists and a large proportion of the technology needed for such a mission is either already available or on the way to being developed. Even so, there is no getting away from the fact that deep space exploration challenges belief. There is no way that we can begin to comprehend the vast distances of space beyond the solar system; no way that we can begin to wrap our minds round the discovery of other forms of civilisation in the dim recesses of the universe. Yet all the statistical analyses argue that they must be there. It would indeed be a most singular accident if the earth were the only body on which sophisticated life forms had evolved.

But deep space exploration is also very costly, and it has a hard time at budget discussions. Most of the money going into space these days is aimed at buying something more tangible; things like navigational accuracy, better communications, military advantage, even better production techniques, because of course space has the singular advantages of purity and freedom from gravity. The first humble workshop in this new era of space exploitation is the space laboratory designed to fit inside the space shuttle.

A New Dawn

Later this year when the space shuttle is scheduled to make its first manned orbital flight, man will have entered a completely new era of space exploration. If the Apollo flights made science fiction, science fact, the shuttle will make space flight an everyday event, almost as normal as catching the 8.15.

For the first time ever, operations in space will become an economic proposition. The massively expensive one-shot spacecraft will become virtually a thing of the past, as the versatile and reusable shuttle becomes the work horse, the delivery van of space. It will be able to carry as much as $29\frac{1}{2}$ metric tonnes of payload in its huge 4.6 by 18.3 metre cargo bay. It will carry satellites and instrument packages up into orbit, but it will also be able to recover payloads and bring them back for repair or maintenance either in space or back on earth. It will be able to seek and retrieve objects in space and reposition them and it will be able to service space laboratories and manned space stations. NASA is already selling freight space in this first aircraft of its intended space fleet.

The shuttle orbiter can be operated by a crew of three although it can carry seven including scientific and technical personnel, and ten in an emergency. Astronauts will fly it there and back, carrying as passengers scientists and engineers with their experiments or their work set up in the cargo hold. In size it is comparable to a modern jet liner: 37 metres long with a wingspan of 24 metres. When you see it taxiing out it looks remarkably ordinary. It is certainly not as striking as Concorde, for example. At launch the orbiter will lift from the pad with its liquid-propellant rocket engines burning in unison with the solid-propellant engines of the two rocket boosters.

The large cylinder directly beneath the orbiter is just a massive fuel tank. It contains 1500 cubic metres of liquid hydrogen in the aft section, and 500 cubic metres of oxygen also in liquid form, in the front part. Both tanks are pressurised to about the same degree as a car tyre. Like the fuel tanks on an aircraft, it feeds its fuel into the main rockets of the orbiter.

The two smaller cylinders either side of this external tank are the solid rocket boosters. Like the orbiter itself these are reusable. They burn for a brief two minutes on take off, consuming 450,000 kilograms of solid rocket fuel in granular form, packed like coarse sand into a hollow tube inside the rocket. About 43 kilometres above the earth these burned out 'boosters separate from the orbiter and swing slowly back to earth slung on giant parachutes. They should land about 300 kilometres down range where a recovery ship will pop a bung into the rocket nozzles and tow them back to port for refurbishing and subsequent reuse. Scientists expect the boosters to be used on at least ten missions.

One of the few throwaway items in the whole assembly is the external tank which will be jettisoned before the space shuttle goes into orbit and in NASA's words 'it will follow a ballistic trajectory and impact in a remote ocean area'. It takes a total time of ten minutes for the space shuttle to go from the launch pad to being in earth orbit.

After the mission, which can last anything up to thirty days, the shuttle will re-enter the atmosphere at about 24,000 kilometres an hour; the world's first truly hypersonic operational aircraft. It will drop from a height of 120 kilometres to sea level in about thirty minutes, subjecting the crew members to very gentle $1\frac{1}{2}$G. Outside, however, things are rather more strenuous. The underside of the shuttle will have to withstand temperatures of 1260°C (2300°F) caused by friction heating as the spacecraft plummets earthwards. The shuttle will land just like a normal earthbound aircraft, using the control surfaces on its wings and rudder. A special runway has been built at the Kennedy Space Center in Florida. It is almost 5 kilometres long to make sure that shuttle crews do not have to worry about running out of space. Two weeks later the shuttle will have been 'turned round', ready to be mated to its external tank and boosters for another trip.

But the space shuttle is only the carrier. The key thing about it is the cargo hold at the back end. They have already planned 570 missions between 1980 and 1991. On each flight it will carry one or more separate payloads and almost anyone can apply for experimental space in the shuttle hold. If you have got $10,000 or so and

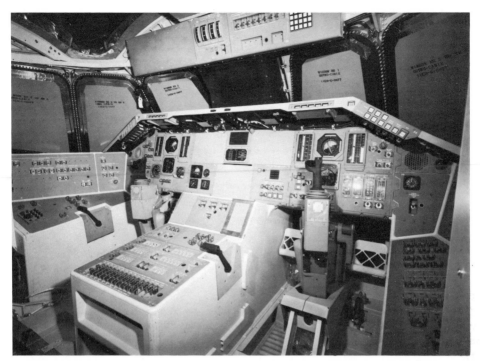

The space shuttle, as much an aircraft as a space vehicle – as the familiar cockpit layout shows

Ion drive probes 'set sail' for a rendezvous with Halley's comet in 1986

your experiment can be fitted into a package no more than 5 cubic feet in volume and weighing less than 90 kilograms, your name could go on the list that fills an armful of telephone-directory-sized books for a place in the space race.

Perhaps the most important service that the shuttle offers will be the placement of satellites in orbit. As many as five at a time can be hoisted into the sky to be placed in orbit by 15-metre-long mechanical arms that can manipulate and position the payload. Each arm has remotely controlled television cameras, and lights on the arm and in the hold will provide illumination while the payload specialists manoeuvre their very special kind of freight into the vacuum of space.

But for the languishing European spacenik 1980 has got to be a date to be reckoned with. For them the shuttle will be used to carry Europe's very own spacelab into orbit.

Europe in Space

Spacelab is a manned laboratory being developed by the European Space Agency in which for the first time European scientists, engineers, and technicians rather than astronauts – women as well as men – will be able to conduct experiments in earth orbit. Mounted in the space shuttle's cargo bay, the first spacelab mission will take off in December 1980. The primary objective of this first flight will be to check out the spacelab's systems and make preliminary measurements of the environment in which the Euronauts will work. These are the people whom NASA calls 'payload specialists'.

The spacelab itself is built a bit like a tube train. The long pressurised module is a totally enclosed environment in which the scientists can work under almost exactly the same conditions as in a laboratory here on earth, except perhaps for the somewhat cramped conditions. The rear section of the spacelab is exposed to the elements. It is a series of pallets – designed in the UK, on which telescopes, monitoring devices, and antennae, are mounted. During the first mission there will be no less than seventy-six separate scientific and technological experiments – sixty European, fifteen American, and one Japanese, all selected by the European Space Agency and NASA. Altogether, scientists from sixteen countries will have some form of direct interest in this first spacelab mission.

Plasma physics, materials processing, biology, botany, medicine, astronomy, solar physics and thermodynamics are just a few of the areas in which experiments will be carried out, and this first mission will only last for about a week. During the daytime the Euronauts will carry out the experiments in the crowded laboratory module and then they will crawl through the tunnel back to the crew's quarters to sleep. In addition to flying the ship astronauts will be responsible for laying on all the other services; power, food, toilet facilities, and so on. It really is taking on all the appearance of a space airline.

Not surprisingly the announcement in 1977 that the European Space Agency was looking for a scientist to be sent into orbit to be one of the two payload specialists (the other will be an American) brought a flood of applications. By October the numbers had been whittled down from over 2000 to fifty-two men and one woman. At the time of writing that list has now been whittled down to four: an Italian, a Dutchman, a German, and a Swiss. Over the next two years they will all go into training – being briefed on the rigours and procedures of a spaceflight, emergency drills, flight plans, and becoming familiarised with zero gravity. Sometime in 1979, one of the four will be named as the first Euronaut.

Apart from all the high-powered research that will be going on as the space shuttle hangs in orbit, Professor Poul Thomsen, of the Royal School of Education Studies in Denmark, has suggested that the spacelab should be used to help in teaching physics to schoolchildren. Not, unfortunately, by taking them up into orbit, but simply by filming a few simple experiments, inexpensive to mount, and easy to perform, that would take advantage of the zero G conditions inside the spacelab. The lack of gravity, for example, means that all objects set in motion within the spacelab will move in a straight line. A tennis ball thrown from one scientist to another will follow a perfectly straight path. Indeed space psychologists report that unlearning the adjustments that we all make unconsciously, all the time, as we move and co-ordinate our actions within the earth's gravity, was one of the most troublesome phases in the training for the Apollo missions.

But apart from having no up or down within the spacelab, Professor Thomsen believes that one would very quickly begin to appreciate the difference between mass and weight. Our ball-playing astronauts would discover that each time they threw the ball they themselves would be nudged in the opposite direction. If they wanted to move a chair, for example, they would soon discover that the chair's centre of mass is not where they expected it to be. Try to pick up a chair as you might in your own sitting-room and you would send yourself spinning across the spacelab because of the laws of conservation of angular momentum.

This relates to another of Professor Thomsen's bright educational ideas. Spin up a medium-sized gyroscope weighing perhaps a kilogram on earth; if the hapless astronaut were to pick up the gyroscope from its stand and hold it in front of him he would slowly start to spin in the same direction as the wheel of the gyroscope, around his centre of mass. By moving the gyroscope he would be able to show that in some positions he was spinning quite quickly, whereas in others he would spin more slowly. So long as he has got a fairly strong stomach it would be easy to do the simple mathematics that showed how angular momentum is always conserved.

To some extent Professor Thomsen's hopes have already been realised since some of the skylab astronauts were filmed gambolling around inside the research module, although the films are not used yet as teaching aids. If Professor Thomsen's plans to extend the range of experiments to include examples of other fundamental laws are successful then perhaps the next generation can look forward to some pretty stimulating astrophysics in its school schedule.

Of course Europe is not only concerned with the spacelab. By far the major part of the ESA's operations consists of building up the European satellite network with satellites for navigation, weather, earth observation, as well as selling their expertise in the growing world market. But there is no doubt that the European effort is still dwarfed by a number of American projects.

Early in this year NASA had hoped to announce its plans to build something so big that as it orbits the earth it will blot out galaxies, planets, and part of the moon, as you watch it pass overhead. Recently President Carter postponed the project until 1983. Its name is 'Powersat' and it is an electricity-generating station which will exist out in space. It is designed to trap the enormous radiant energy of the sun and convert it to a thin microwave beam that could be directed towards receivers here on earth. The Powersat – and they are contemplating more than one – will stretch almost 24 kilometres through space like a vast free-floating kite. It will be made up of four separate collecting dishes, each 4 kilometres across. By the early 2100s there could be no less than thirty of them hanging motionless over fixed points on earth, beaming 10,000 million watts down to us every year. The aerospace engineers who have been proposing various Powersat schemes since the late 1960s are convinced that an orbiting power station of this type would pay for itself within a decade. By their calculation within forty years the energy from one station would be worth $62 billion a year. It is claimed that electricity could be supplied to homes and industry at about two-thirds the cost of present methods of power generation.

Each of the separate facets in the 4-kilometre-wide dish would be 10 metres across and steered individually to focus the sun's rays on the heat generator high above it. The heat generator works rather like a radiator in reverse, absorbing heat rather than giving it out. In one of the designs, the concentrated solar energy enters an absorption panel where it falls upon kilometre after kilometre of tightly-wound, thin, black tubing filled with liquid helium. The expanding helium drives a turbine and a compressor, which turn the electricity generator. The electric power that is generated is then converted into radio energy and beamed out from a transmitter at one end of the satellite array. The transmitter alone is 1 kilometre in diameter, and by the time the beam reaches earth it covers an area 5 miles across. No less than 95 per cent of the transmitted energy would get through to the surface of the earth.

The ground system – the power receiver – will take the form of row upon row of angled panels of microwave sensors. They will convert the 50 megawatts falling on every square centimetre back into electric power for domestic use. The energy output each year from a single Powersat unit of this size would be so great that

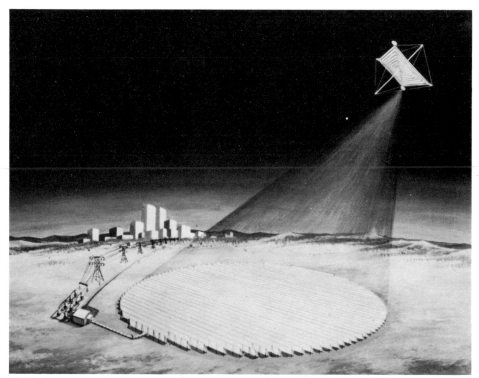

Fields of receivers will collect energy from the power stations in the sky during the day and night

A power satellite begins to take shape in orbit

within nine months the engineers estimate that it will have broken even in energy terms, that is, given back the energy required to build and launch it.

The problem with Powersat is not so much its design, or its efficiency once in position; it is simply getting that amount of hardware into orbit out in space. It would require one flight every day for a year from a launch vehicle larger than *Saturn V*. Powersat, and projects like it, require a revolution in space-freight techniques. But already that revolution is being created on the drawing-boards of some of the big American aerospace companies. They have already, for example, designed a monster called LEO; LEO for Low Earth Orbit vehicle. It stands almost 60 metres high and it is 48 metres across. That is about as big as a twenty-storey block of flats. The cargo is carried in the nose – inside a smaller rocket called a 'kicker'. It is a bit like a Russian doll in concept, because inside the 'kicker' rocket are the smaller modules that are used in constructing the satellite. The whole of the rest of the body is a fuel tank containing liquid hydrogen, and liquid oxygen.

But the really interesting thing about LEO is that it will take off and land on water. According to an artist's impression of Plot 39 at the Kennedy Space Center, when the Powers at launch and recovery basins have been made, each circular pond is 8 kilometres across. When LEO has been assembled it will be

LEO, the heavy cargo Low Earth Orbit freighter, heads for splash-down

towed down a water-filled canal which will serve as the launch pad. After lifting off and delivering the cargo rocket into orbit LEO will re-enter the earth's atmosphere heading for splash-down just off the coast of Florida. Once it has been established that LEO has re-entered safely, the flight path will be altered slightly so that the rocket can land in the fresh water recovery basin at 450 metres altitude and the retrorockets will be fired to lower the huge structure gently on to the water.

A project like Powersat and LEO would involve the mobilisation of space technology on a scale that would make the Apollo programme look like a Fourth of July picnic. Satellites the size of cities and rockets as big as towerblocks may seem far-fetched. However, the difficulty is not so much technological as psychological. One of the managers of an American design team working on space power programmes calls it 'concept shock'. But even a concept as colossal as Powersat pales somewhat when compared with another programme that has, in a sense, been on the stocks for over 100 years – project 'Brick Moon'.

It was as long ago as 1869 that a far-sighted novelist, Edward Everett Hale, wrote a story called *The Brick Moon* about the accidental creation of a space colony. A brick sphere intended for guiding maritime navigators was to be catapulted into earth orbit by rotating wheels. This in itself is uncannily prophetic. The European Space Agency started work in 1974 on a maritime satellite called Marots, designed to do more or less the same thing. At the end of 1978 Marots took its place alongside three Marisat satellites launched the previous year and already helping to guide ships more accurately along the sea-lanes.

But back to Edward Everett Hale. During the construction of this brick sphere something went terribly wrong. Still containing many workers inside, it rolled on to the catapult too soon and the first space colony was launched. Fortunately the workers had plenty of food and supplies and there were even a few hens which had strayed into the sphere; they decided to sit back and enjoy their fate.

One hundred years later in 1969 at Princeton University there was a detailed study of the practical possibilities of space colonisation. The result was pretty conclusive. It could be done. After a ten-week seminar held by NASA in the summer of 1975 'could' in some people's view became 'should', as it appeared that the sytem would pay for itself within a generation, and economic arguments play a critical role in all the studies. What is being proposed is not in any way someone's fantasy about space colonisation. It is being put forward as a commercial proposition.

The key to the colony's success, it is argued, would lie in the development of the twin resources of abundant solar energy and cheap minerals recovered from the moon. Not only does the uninterrupted sunshine of space offer high productivity for farming, but it also provides ample energy for a whole range of industries. Using solar energy to generate electricity and to power solar furnaces, the colonists will begin to carve up the moon, towing giant ore barges out to the industrial space islands to refine metals such as aluminium, titanium, and silicon. Apart from being valuable, even critical, for earth-based industry, these are exactly the materials that would be needed to manufacture the space-based solar power stations we looked at earlier. So, the space colony would not only justify its own existence, but it would also provide a commercial return. Looking further into the future, all kinds of industrial processes that are either dangerous, polluting, or could benefit from reduced or zero gravity, might be located in space. Some people have suggested that space islands might also contribute to alleviating the world's population problems, but that scarcely seems likely. Optimistic estimates of the global population suggest that there will be 10 billion of us in the year 2035. An odd 10,000 or so (the likely population of a second-generation space colony) is not even going to be noticed.

It has been suggested that such a colony could be well under way within fifteen years although one should perhaps bear in mind a comment made many years ago by a great futurologist, Arthur C Clarke. In his view, people who attempt to look into the future tend to be too optimistic in the short run, and too pessimistic in the long run; too optimistic because they tend to underestimate the forces of inertia, which can delay the acceptance of new ideas; too pessimistic because once acceptance has been won, development tends to follow an accelerating exponential curve.

But that having been said there are already a number of design studies in an advanced state of preparation. There are about half a dozen fundamental design requirements that have to be met before you can start working on all the rest:
- The space colony must have an atmosphere, so that its occupants can live and work at least most of the time in normal earth-like 'shirt-sleeve' conditions. This means that it must be enclosed.
- The settlers must be protected from radiation.
- They must be provided with a regular earth-like cycle of night and day.
- The space colony must have a gravity of approximately 1G, not simply to keep the colonists on their feet, but to prevent the deterioration in muscle and bone tissue that occurs rapidly under conditions of weightlessness. This means that it must revolve constantly around its axis.
- The space colony must also be attractive, a pleasant environment in which to live and work, otherwise nobody is going to settle there.
- Finally, it must be self-sufficient in food and materials. It would be impossible for a settlement of 10,000 people to receive regular supplies from earth.

Even at this simple level of analysis the design options are immediately restricted. Four fundamental shapes present themselves: a sphere, a cylinder, a dumb-bell, and a doughnut, or as the mathematicians describe it, a torus. Current thinking in space settlement is inclined toward a combination of a sphere linked to a number of torus-shaped rings. This space settlement is called a Bernal sphere configuration in honour of J D Bernal. It consists of a central spherical habitat almost a mile in diameter, placed between two clusters of toruses which will house the agricultural areas of the colony. The Bernal sphere offers a number of key advantages and it is particularly efficient in protecting the inhabitants from harmful radiation. It is immensely strong and yet comparatively simple to construct; its gravitational requirements are relatively straightforward. The main disadvantage appears to be that it requires three to four times as much atmosphere mass as other possible configurations.

Food will be grown on the farms located in the toruses at either end. There is no reason why all the normal range of arable crops and farm animals should not be grown and reared there, and in an environment that is germ-free and has a totally controlled climate, the colonists could look forward to startlingly increased rates of production. In fact it would be something of a farmers' paradise; no pests, no storms, no late springs nor early frosts to frustrate their efforts. But what would it be like to live on – this manufactured space island?

Well, the available surface area for each man, woman, and child, would be around about 40 square metres. That may not sound very much but it's comparable with the sort of space taken up by many urban residential areas here on earth: Rome or New York for example. Of course it is just surface area, there's nothing to stop your building a three-storey house in the colony.

In one design for a second-generation space island the idea is to recreate an

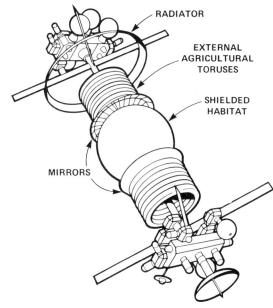

RADIATOR

EXTERNAL
AGRICULTURAL
TORUSES

SHIELDED
HABITAT

MIRRORS

A schematic diagram of a space colony built round the notion of a Bernal sphere

earth-style landscape, perhaps even modelling it on one or more highly regarded earthly beauty spots. The landscape would include villages separated by open spaces and parklands, streams and low hills, woods and fields. There is no reason why the space settlers should not carry on very much the same range of activities and sports as they do on earth and a slightly lower gravity could do wonders for human athletic endeavour.

Sunlight would enter the colony through being reflected from a vast system of mirrors at either end of the sphere, which would be steered to provide a cycle of night and day.

Outside the sphere a shield of lunar material would be used to protect the humans inside from the radiations of space and to absorb the impact of the occasional meteoroid. The chances of a really large asteroid smashing into the colony are so remote that for all practical purposes they can be ignored. The frequency, for example, of a meteoroid weighing as much as an apple striking any part of the structure is something like once every 200 years.

So having provided this home from home, with a population, a source of income, and all the recreational amenities of twenty-first-century life, where are we going to put it?

At certain points between the earth and the moon there are 'dimples' in the gravitational fields caused by the interaction of these two planetary bodies. Places, in fact, where the gravitational fields of the earth and the moon cancel each other out. They are known as the Lagrangian libration points. There are five in all, and they have the useful property of stagnation. Anything put there will stay there. The three Lagrangian points directly in line with the moon and earth are in a sense saddle-shaped and hence not quite as stable as the fourth and fifth Lagrangian points which are bowl-shaped. These are the two points at which the colony, or space colonies, since whole archipelagos of them are envisaged, could be moored in complete safety. The earth and the moon would be about 400,000 kilometres away.

Although the earth would of course still represent 'home', by far the most important nearby planetary body would be the moon. Over a million tons of lunar material could be mined from the surface of the moon every year and shipped from the moon in its raw state to the colonies' industrial zones. Because the moon's gravity well is twenty-two times shallower than the earth's it is much easier to mine vast loads of ore there. Bearing this fact in mind, a very ingenious method of launching the crude ore towards the colony has been proposed. Stretching out across the lunar landscape would be a 10-kilometre-long track of electromagnetic hoops. The ore would be loaded into large containers or buckets. Built into the walls of each bucket there would be liquid-helium-cooled, super-conducting magnets to suspend it above the track. Accelerated up to 2400 metres a second by the giant linear electric motor, the buckets will experience 30G for their five-second journey across the lunar landscape. The bucket would be constrained by the track to follow the curve of the lunar surface. The payload would not. So it would be hurled into space. The bucket would then enter a 3-kilometre stretch of track where another set of linear motors would bring it to a halt, while the payload would coast onwards to its destination.

A colony which consists of a habitable sphere a mile in diameter represents no more than the starting point of a major space colonisation programme. It would be a prototype; a place for trying out concepts and resolving design issues. But already there are space scientists working on schemes for giant cylindrical space stations 20 kilometres long and 5 kilometres across, ringed by industrial, agricultural and power-generating satellites and complete with commuter transports to carry the work force to and fro. By then the colonies would be harvesting their ores from way out in the asteroid belt, as well as the moon.

The cost of a prototype colony is in the region of 25 billion dollars – a mere fraction of what America spends on armaments in a year. The technology needed could be available in ten or fifteen years. The great lack at the moment is motivation. The argument is that when the Western world reaches the dreaded power gap that is forecast for the mid 1990s, the economics of space colonies and their attendant solar power stations will become very attractive. If the scientists are right then there are likely to be children reading this, whose own children could be born in a space satellite – poised indefinitely between the earth and the moon.

A Sea of Tranquillity?

In the autumn of 1976, a Pentagon spy satellite nicknamed the *Big Bird* was drifting up above the Indian Ocean minding somebody else's business when suddenly and unexpectedly a powerful beam of infra-red light blinded it.

It had been searching the Central Russian flatlands for heat traces that would betray the location of Soviet missile bases. It was put out of action for ten days during which time the Americans had no idea what was going on in the Siberian wastes below. A great deal of hard thinking went on and eventually an official statement from the American Government put the cause of *Big Bird*'s disability down to the heat traces given off by the natural gas flares in the oilfields of central Russia. Not everyone is entirely satisfied with this explanation and some have even suggested that it would be impossible to blind *Big Bird* in this way.

It sounds like the scenario for another of the science fiction films that have been so successful recently, but it is not. Since the mid-1960s both Russia and America have included in their space research programmes a substantial amount of what one might call military activity. Back in 1967, after the two Super Powers ratified a treaty outlawing nuclear weapons in space, most people assumed that the heavens had been prohibited from being used as a battlefield. In the same year, however, the Soviet Union was conducting tests with the Cosmos series of satellites; this has been widely interpreted as research into the feasibility of developing killer satellites. It appears that it was sending up pairs of Cosmos satellites, one as the target, and the other as the killer. The general idea was to manoeuvre the killer alongside the target to disable it. Out of the twenty-seven such Cosmos launches, seven have ended up with the killer satellite exploding. This is probably not the most effective way of destroying target satellites since, in the near vacuum of space, shock waves have little effect. Unless the killer satellite were extremely close it would be unlikely that blast or flying debris would do any damage. Radiation is much more likely as a destructive weapon; high energy lasers or ion beams directed from the killer satellite to its target. You do not have to go to the trouble of blowing it to bits if making it blow a fuse will switch the target satellite off just as effectively.

Recently, perhaps in response to the development by the Chinese of their own reconnaissance satellite, the Russian killer satellite programme has been increased.

For their part, in September 1977, the United States Air Force announced the award of a $58\frac{1}{2}$-million-dollar contract for continued development of a killer satellite. The programme had actually been initiated in 1975, but as long ago as 1963 the Americans were looking at the possibility of destroying satellites with missiles. A modified Thor missile was fired towards a booster rocket in low earth orbit which was acting as the target. The missile was brought within range for a simulated nuclear kill, that being the most effective way of saturating an enemy satellite with lethal radiation. When the project was brought to an end by the 1967 treaty, a number of further tests were conducted using conventional warheads, but the programme was a flop and was finally abandoned.

As far as the Super Powers are concerned, satellite killing poses a major threat. Present-day military operations depend heavily on satellites for intelligence-gathering, navigation, weather information, command and control, and communications. Everything from requisitions for more toilet paper to ballistic missile guidance uses satellite links. One Pentagon spokesman has suggested that it would take no more than a week to put every satellite the Americans have out of action. That is one good reason why $58\frac{1}{2}$ million have just been tipped into the killer satellite coffers.

In May 1977, an argument flared up in the United States concerning the objectives of a mysterious installation in Kazakhstan in central Russia. A major-general in the United States Air Force raised the question of whether or not the Soviets were developing a ground-based system for destroying satellites in addition to the Cosmos programme. He was worried by what Air Force Intelligence circles called P-NUTS (Possible Nuclear Underground Test Site). His contention was that the Soviets were perfecting a technique for containing the power of a fusion explosion and converting it into pulses of high-energy-charged particles which could be beamed towards any object in orbit, destroying it instantaneously. In support of this theory, on seven occasions since 1975, spy satellites snooping around in the area have found traces of tritium and hydrogen above the test site. These could have been the products of a fusion explosion.

The mystery deepened still further when a Soviet academician, Leonid I Rudakov, visited a scientific seminar in the United States to deliver a paper on high energy particle physics and fusion reactions. The story goes that even before the polite applause had died away, scientists at the seminar were being warned by intelligence experts who swore them all to secrecy, confiscated their notes, and the blackboard. It would appear that the Soviets are significantly further ahead in this field than had been realised. One wonders what happened to Rudakov when he got home. The debate concerning the mysterious goings-on in Kazakhstan is still unresolved. Some scientists have suggested that if the Soviets had actually got the technical ability and the financial resources to build a charge particle beam weapon, that is not what they would have built when high energy lasers offer a simpler alternative. Others claim that it would be impossible to aim such a device accurately enough because of the interaction of the earth's magnetic field with the particles in the beam.

Meanwhile the US Defense Department's own beam weapon research leads it to believe that the problems inherent in the construction of such a weapon could not be solved within a decade. The chilling rebuff to that argument, offered by those intelligence specialists who believe in the beam weapon theory, is that while that may well be true for the Americans, Soviet technology in this area is ten years ahead. I, for one, am hopeful that it remains a mystery for many many years to come.

Finally, we must not forget the shuttle. Its uses as a cheap manned surveillance platform are obvious, as are its abilities to move around releasing satellites that are non-transmitting, and therefore not detectable – satellites that can loiter in orbit waiting to be triggered into activity when the need arises. Not so obvious perhaps is the fact that an entire Soviet Salyut space station would fit comfortably into the shuttle's hold. The shuttle's manipulator arms could also be used to tear the solar panels from an enemy's reconnaissance satellite – and then swing it into its cargo bay, never to be seen again.

Space is beginning to look less and less like the peaceful place we invaded twenty years ago. Dr Malcolm Curry, the Director of Research at the Pentagon, declared in a recent speech: 'The Soviets have seized the initiative in an area which we hoped would be left untapped. They have opened the sector of space as a new dimension for warfare, with all that implies. I would warn them that they have started down a dangerous road.' You cannot mix more metaphors than that but let us hope that war among the stars is a 'dangerous road' which both the Americans and the Russians will leave unexplored.

The New Astronomy

By the end of this century the *Voyagers* will have unlocked many secrets concerning the origins of the solar system. Mars Rovers may have answered the question as to whether there is life on Mars. And men may be installing themselves in moon bases and space stations preparing the way for the colonists who will follow. How will our understanding of the mechanics of the universe have changed by then? Over the next five or six years we shall see the launching of a whole new breed of astronomical satellites; astronomical, that is, in purpose as well as in cost.

The edge of the observable universe is probably about 15,000 million light years away. Over the past few years advances in astronomy have enabled scientists to push the range of their observations further and further out into deep space; yet telescopes on earth are limited by the blanket of the earth's atmosphere.

In late 1983 NASA plans to launch, from the shuttle, a powerful new eye in the sky; an optical telescope that will enable astronomers to look seven times deeper into space than ever before. Circling approximately 380 miles above the earth, its $2\frac{1}{2}$-metre diameter optics will be free from atmospheric interference. It will be able to lock on to its targets with absolute accuracy and track them for periods as long as thirty to forty hours at a time, picking up objects fifty times fainter than those which it has previously been possible to see. To do all this requires a directional stability of .007 seconds arc during the observation. That is roughly equal to a marksman in Edinburgh zeroing in on a penny-sized target in Southampton. Little wonder that astronomers see the launching of the large space telescope as one of the greatest astronomical events of the century. It will enable them to probe deeper than ever before into those age-old mysteries that surround the origin and structure and evolution of the universe. They will also be able to detect planets circling in other solar systems; new worlds seen for the first time.

The telescope will be maintained on station while it is in orbit by regular visits from the shuttle crews and every now and then, when necessary, it can be returned to earth for major refurbishing and adjustment. But star gazing of this quality does not come cheaply. NASA's contract with Lockheed and Perkin-Elmer, the subcontractors building the telescope, is expected to come to $131 million.

Towards the end of 1970, an American satellite took off from an Italian launch pad just off the coast of East Africa. Launched from this location so that the spin of the earth could give the satellite the most effective possible kick into space, it was the first satellite wholly dedicated to X ray astronomy. Codenamed Uhuru, the

Swahili word for 'Freedom', it focused astronomers' attention on a theory that had been gathering dust for over a century and a half. Briefly, the theory states that there may be regions in space where gravity is so strong that no radiation, not even light, can escape from them, so they cannot possibly be seen from earth. These regions are now called 'black holes'. An optically visible star suddenly turned out to be a strong X ray source as well. Now, visible stars cannot themselves by sources of hard X rays because far higher temperatures than those generated on ordinary stars are needed to produce hard X rays; one of the few mechanisms that could produce hard X rays in profusion would be the attraction of gas into a very strong gravitational field. So scientists wondered whether the X ray source Cygnus X-1 could be a black hole alongside its visible super giant companion star, swallowing vast amounts of matter from it and hence releasing X rays. In 1973 many astronomers felt that this was indeed what was happening and that a black hole had at last been identified. Unfortunately, since then other explanations for the phenomenon have been offered and so we cannot yet say for certain whether or not the mathematically feasible black holes really do exist.

In 1980 the European Space Agency is due to launch EXOSAT which will continue to examine the structure of some of these still unexplained X ray sources. To do this it will have to be launched into a rather peculiar orbit which takes it well out of the plane in which the moon orbits the earth. The reason for this is that from the satellite's point of view the moon will appear to swing all over the sky. By working out the exact time at which the moon blots out X ray sources in the heavens and by computing and correlating the position of the three points of the triangle, the earth, the moon, and the satellite, the position of the X ray source can be pinpointed. Thanks to EXOSAT's peculiar orbit it will be possible to apply this blotting-out technique, known as lunar occultation, to most of the known X ray sources. Of particular interest will be star systems like Cygnus, and one of EXOSAT's aims will be to pin down the location of those elusive black holes.

In August 1977, a mission was launched from the Kennedy Space Center which might just pip EXOSAT at the post in the race to define the black hole. The first of three planned missions was called the High Energy Astronomy Observatory (HEAO-A) and this was joined in space by HEAO-B and HEAO-C in 1978 and in 1979 respectively. These observatories, the heaviest unmanned earth orbiting satellites ever launched by the United States, carry a battery of scientific instruments capable of detecting X rays, gamma rays, and cosmic rays. Rotating end-over-end the HEAO-A spacecraft will survey the entire sky every six months. During the year-long mission the three space craft have each been allocated separate tasks. But a key objective of all these missions is to learn more about the way in which X rays, with a thousand times the energy of ordinary light, and gamma rays, millions of times as energetic as ordinary light, are sent from deep space towards the earth.

Space has been described by one astronomer as an immensely powerful laboratory. Theoreticians and researchers back on earth, working in fields such as high-energy particle physics, have all the advantages of control over the conditions of their experiments. But in space, events, experiments if you like, of unimaginable scope and violence are taking place all the time. Space is the ultimate research laboratory. Until comparatively recently man's observation of these events has been clouded by the veil of the earth's atmosphere. Astronomy from outer space lifts that veil. We are just on the threshold of a clearer view than ever before of what lies out in the wild black yonder. So it looks as if the next decade is going to be a period of great excitement and achievement for a highly privileged and generally very small body of men: astronomers and space scientists. But what is in it for the rest of us? It is a question that even now, twenty years into the space age, still evokes a remarkable range of responses.

One of the commonest perhaps starts off with emphasising the value of non-stick frying pans and fireproof underwear for racing drivers, having gone through a list of hardware big and small that you can buy over the counter.

Another, just as common, begins with a somewhat glazed expression in the eyes and talk of 'the question missing the whole point of the endeavour' and follows up with the richness of the space programme in terms of adding to man's fundamental knowledge; filling in the gaps in his understanding of the cosmic jigsaw puzzle. Both are right of course. There is no single complete answer. Neither is there much doubt that one of the reasons why space is being explored is because it is there, like Everest. And at this point in man's history a small handful of nations have the wealth and the skills to take on space exploration.

In the mad days of the 1950s, and the all-out race for the moon and anything else in sight, the objective was simply to get there. The moon race itself had surprisingly little to do with the dominant use for the biggest benefit that has so far emerged from space technology, namely providing a platform for looking back at the earth's rocks, and seas, and weather and vegetation and military bases with every possible kind of eye. Ordinary cameras only pick up the same information as the human eye, immensely valuable but nonetheless limited to a particular fraction of the entire spectrum of electromagnetic radiation that carries with it information. A great deal of work over the past ten years has gone into the development of devices more efficient at picking up other wavelengths in order to build up a multi-layered, multi-coloured, infinitely more detailed picture of the planet spinning beneath, and of what is happening in the slim envelope of gas that lies above its surface.

The past twenty years of space exploration have been a phenomenal success story. It would be difficult to match them in any other field of human endeavour. But perhaps the single element of space spin-off that has been most ignored, and undervalued, possibly the one that could still come to have the greatest effect on the stability and development of our civilisation in the years to come, has been the numbing sense of perspective that space travel brings. To a man, all astronauts, whatever their nationality, seem to have experienced it, at once astonishing and intensely humbling, as they gazed out of their spaceship window to look back at spaceship earth.

Michael Rodd

Leisure is a Serious Business

If I had been asked to write this chapter three years ago I would have got it all wrong. I would have been forecasting with great certainty that in the last couple of years of the 1970s we would be spending our leisure hours listening to music at home reproduced to sound as if we were in the concert hall or the studio in which it was originally recorded.

Quadraphonic or surround sound was the coming phenomenon in the leisure business. It had a great deal to commend it.

The weakness of a single channel or even a stereo two channel sound system is that the effect tends to be flat. The sound stage may have width but it has little depth and all recordings are made in 3-dimensional rather than 2-dimensional circumstances. Four channels rather than two seemed a logical development, with the extra channels providing the ambience and the reverberations which would give our reproduction the true atmosphere of the original.

So why would I have been wrong? Because the quadraphonic revolution has not happened. That is not to say it will never happen, but it has not yet, and to analyse why might be a salutary exercise before I embark on what I now think the years to come have to offer us in our leisure hours.

The reason surround sound, and let me settle on that description, has not had an easy time of it is that the manufacturers have not agreed on what it is. It is hard enough trying to sell a new idea to anyone, but if those who are doing the selling all give different versions of the story, no one is going to find the idea itself easy to grasp.

In seeking to ensure that four-channel discs would also reproduce accurately on existing two-channel equipment, the record industry came to a number of different conclusions. Some manufactured matrix encoders to process the four separate channels into two for disc recording. These two channels generated a stereo image, or if reprocessed through a suitable decoder, regenerated the original four channels. But there was no agreement on the matrix. Some record labels processed their surround sound using one system, others used another and then there were those who did not produce surround sound discs at all.

Then another batch of record labels used a quite different system. The disc actually carries all four channels; two are conventional stereo, the other two are on a high-frequency carrier wave which is detected on playback and used to sort out the four channels.

Yet another camp suggested using just two channels for the recorded information, but including in those two channels such information as height of sound source and distance from listener. Complicated mathematics were needed to produce the two channels, and an interesting microphone array had to be perfected which picked up the sound as the human ear does.

Broadcasting stations could not agree on a standard. The BBC, watched from all over the world as leader in broadcasting equipment research and production, was developing its own format called Matrix H. Other stations not prepared to wait for a move from London split their allegiance between the two other matrix formats already adopted by some record companies.

More recently the BBC has been working with a team from the University of Reading who have developed yet another surround-sound technique. Known as Ambisonics, it owes much to the work of Oxford mathematician Michael Gerzon. The two organisations have now produced what they call System HJ. This could become a widely accepted standard for broadcast and possibly also recorded surround sound.

It is hardly surprising, looking back, that not many customers paid out for equipment with which to enjoy the new surround sound. What were we supposed to buy? And yet so much work was being put into the new development at the time that it was hard to see it not coming to fruition somehow before now.

I say all this because this desire of manufacturers to develop independent technologies bedevils the home electronics industry in more than just the field of surround sound. As you read this chapter you will find it crops up again, and though I make no early apology for what follows, the job of making a rock-solid forecast of leisure in the future is a tough one.

However, let me start with an inventor and manufacturer of refreshingly independent outlook.

Photography

Few editions of *Tomorrow's World* are without fascinating pieces of technology but one of the most fascinating I can remember came on the evening we demonstrated the latest achievement of Dr Edwin Land.

Instant photography was nothing new. Dr Land had demonstrated his first system almost thirty years earlier but on this occasion we saw for the first time a full *colour* photograph which developed before our eyes under the lights of the studio. To someone brought up to believe that the magic of photography took place in the dark, and who had spent many hours as a boy proving it in the darkroom, this was really something.

Since then Dr Land has gone on to develop his first instant movie system and has found that his major rival in the home photography business, Kodak, has developed its own competing instant still system. Later in this chapter we'll see

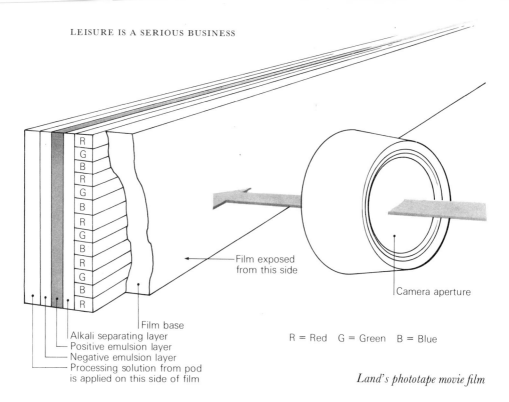

Film exposed
from this side

Camera aperture

Film base
Alkali separating layer
Positive emulsion layer
Negative emulsion layer
Processing solution from pod
is applied on this side of film

R = Red   G = Green   B = Blue

Land's phototape movie film

why Dr Land may have misjudged the market with his movie product, but this leisure technology certainly deserves our attention. First, the new still photographs.

Both Dr Land and the Kodak development team have come up with a complex chemical 'sandwich' which serves three functions. The first is to record the image when the photograph is taken, the second is to build a chemical darkroom round that recording as the film is ejected from the camera, and the third is to turn that record into a full colour image.

There are differences between the two systems. Kodak instant pictures are exposed and viewed from different sides; Land's film is exposed and viewed from the same side. Kodak opted for a reversal film. In other words, the light-sensitive layers which receive the original images when the photograph is taken are later processed into the finished picture. Dr Land has adopted a two-stage system with an original negative leading by diffusion to a positive image when processing takes place. The positive image becomes visible whilst the negative is hidden behind a white-base layer in the chemical sandwich.

But without wishing to deny patent lawyers, or anyone else, scope for argument, let me concentrate on the similiarities in the two systems because they illustrate some fascinating photographic ideas.

Both systems depend on the basis of all colour photography, the principle put forward by Sir Isaac Newton in 1665. That is, white light can be broken down into separate colours, and if those colours are recombined, white light is produced once more.

So both chemical sandwiches have three separate light-sensitive layers. One to pick up red light, one to pick up green and the third for blue. Naturally any intervening colour will be picked up by more than one layer.

Each colour layer is impregnated with the light-sensitive chemical first used by the Victorian photographers, silver halide. So a red signal will expose the silver halide in the red layer, but not the blue or green layers; a white signal will expose the silver halide in all three, but a black signal will expose none. When a photograph is taken the light passes to the various sensitive layers, the appropriate grains of silver halide are exposed and there for the moment the matter rests.

But it is only a brief moment because in both systems a motor immediately ejects the exposed sandwich from the camera, breaking as it does the pod of processing chemicals which immediately produce the chemical darkroom and start the picture development.

This identifies where the silver has or has not been exposed and uses that pattern of exposure to release coloured dyes. The position of the dyes and the proportion in which they are released are the keys to the finished print, providing the colour and the definition of the final picture. As the image forms, the front of the chemical darkroom breaks down and the picture appears before our eyes.

Although a colour system, the amazing thing is that the instant movie film (or 'phototape' as it is called) is in fact a black and white film.

How can a black and white film produce a coloured image? We can recall the original principle that white light can be broken down into separate colours and then recombined; Dr Land's phototape has on it an ultra fine series of red, green and blue lines – 4500 of them to the inch. Red light will only expose the film behind the red filter line and so on for green and blue, although the lines are much too fine to be resolved by the human eye.

The phototape, which is contained in a cassette, is exposed in a movie camera in a normal manner. The cassette is then placed in a player unit which resembles a small portable TV. As the film is rewound in the cassette, a microscopically thin layer of activating agent is spread over the film, developing the previously exposed areas behind the fine red, green and blue lines. The negative image is very weak, but at the same time a very strong positive image is formed.

As a result, when the film is projected, light emerging from the phototape bears the same colour fidelity as the light originally recorded. In other words, if we consider a red exposure then a very weak, almost invisible black and white 'negative' image is formed behind the blue and green lines. So when a light is shone through the developed film from the projector lamp, no light will pass through blue and green areas, but it will pass through red exposure areas to reform the original red light.

The camera, phototape and viewer of Dr Land's system

This principle of red, green and blue lines was used in some of the earliest colour film, but the difficulty originally was getting the fine lines on the film. What Dr Land has done is to produce clarity and remarkable colour quality on film just 8 mm wide by inventing a process to put 1800 red, green and blue lines in every centimetre of film.

It has taken Land thirty years to develop his instant movies and his perseverance and inventive brilliance are worthy of anyone's congratulations. Land's system takes fine pictures, produces fast results and is on the verge of adding sound to its images but because it is inside a cassette, editing and reassembling the finished film are exceedingly difficult. What makes life tricky for Dr Land is that there is now available an even faster system for taking pictures, which already offers sound and editing ability. It is called 'home video'.

Home Television

Much newer than photography, the home video revolution centres on the development of machines which can record television pictures as well as sound signals on magnetic tape. Broadcast studios and other professional programme makers have been using video tape recorders for many years, but it is only in very recent times that the prospect of domestic video recording has become certain.

The transition from professional standard equipment to units which domestic buyers would be able to deal with was almost certainly made by the Japanese in the late 1960s. They were working on smaller versions of the professional reel-to-reel video tape recorders. But these were mostly black and white units and were often far from straightforward to handle. Philips, the European electronics firm based in Eindhoven, Holland, was one of the first manufacturers to realise that if domestic video was ever to succeed, a format had to be developed which was much easier to deal with than reel-to-reel equipment – and it had to offer colour pictures. Potential customers would not be lured from their high-quality broadcast programmes by poor-quality black and white pictures coming from complicated equipment.

Philips' research and development came up with VCR. Like the audio cassette before it, incidentally another Philips' development, the video cassette recording is an attempt to meet the two criteria. Colour it certainly manages; whether using it is straightforward enough for the mass market remains to be seen.

Not that Philips' main rivals, the Japanese, have found that second problem any easier to overcome. Sony also developed a cassette system called U-matic. Its equipment has just as many controls and meters as the VCR units. The two formats, though they both use magnetic tape in a self-contained cassette, are incompatible with one another. It seems that the Japanese, who in the main have been happy to accept the Philips audio cassette as a standard design, have not been prepared to allow the Europeans to call the shots in the video cassette field as well. That failure to standardise, so common now in the domestic entertainment industry, is going to be a major difficulty in developing the video market.

Initially the problem was not a serious one. Philips priced their VCR within striking distance of affluent home buyers, Sony priced the U-Matic safely outside all but the most rich. The result was that U-matic became a standard in industrial video. Sony is used widely in the making of closed-circuit programmes for training and education, marketing and sales promotion. Philips' VCR, on the other hand, has won the bulk of what domestic market there was, together with the industrial market which did not want to risk the higher outlay for the Japanese system.

The day of the domestic video boom may still be a long way off. Two major hurdles have to be overcome, assuming of course that all the potential buyers can be persuaded that recording sound and picture is something that an amateur can do. The first was the price of the units and the second the duration of the cassettes. Even if most of us can afford a video recorder the maximum recording time available of around one hour is somewhat short.

Recently a number of things have happened to push the home video business forward. Philips has developed a half-speed variation on their original design which can record for two-and-a-half hours on one cassette. The Japanese electronics firm, Matsushita, has produced a new cassette format, different again from U-Matic and VCR, called VHS, which manages three hours without changing tapes. The final new influence on the domestic market comes from Sony who have produced yet another format known as Betamax. This time Sony is aiming straight at the domestic market and with a three-hour capability.

The camps are forming. Philips has its followers, Sony and Matsushita have theirs, but the result can only be a drastic cutting of prices as the manufacturers fight for a viable share of the market. It is often said that computers and calculators are the only things to have dropped in price since 1973. Watch out for home video equipment to do the same thing.

And programmes to watch on it

Once we have our equipment how are we going to use it? Herein lies the problem which may prove very difficult to solve because all television programmes transmitted over the air are the copyright of the company who made them.

Anyone who has a television set and a licence to use it can watch those programmes, but recording them is another matter. Like the struggle at the moment to establish some workable licensing authority for the copying of commercial audio records on to tape for private use, it may be that enthusiastic viewers will buy licences for recording television programmes. Until they do, the danger of becoming the first *Tomorrow's World* viewer to be prosecuted for

A video disc under test, showing the laser underneath

A Japanese system for video cassettes – the U-Matic

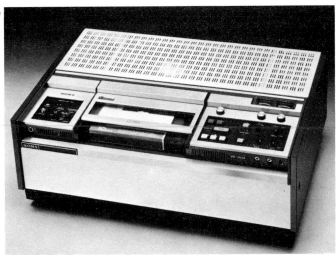

The VLP from Eindhoven – uses discs, not tapes, and cannot record

copying your favourite programme on to tape will, theoretically, hang over you.

The other possibility is that we will use our home video in association with a new generation of small television cameras with built-in microphones. This will give us direct access to truly instant home movies, because tape does not need even the few seconds of Dr Land's photographic miracle. With sound also on tape, and the ability to edit sequences together already available, this application of home video will also help the market grow.

On top of all this comes the long-promised video disc. This is likely, in one form or another, to be with us before 1980. The VLP, which I first reported on from Eindhoven in 1974, is a quite different beast from any cassette recording system.

To start with, the disc does not record anything. Like the gramophone before it, it simply reproduces a programme already encoded into the plastic. The video disc is 12 inches in diameter, and single-sided. A small laser beams up to the underside of the disc and the reflected pattern generated by tiny troughs in the disc is picked up. This pattern becomes the sound and picture of the disc programme. Unlike an audio disc, the video disc can be played backwards, can produce still frames, and can repeat action in slow motion.

The picture quality, and one cannot say this of all the domestic cassette systems, is superb. Of course, one should not overlook the efforts of our own Decca organisation which, in league with Telefunken of Germany, has had a video disc system on the market in parts of Europe for the last three years or more. That system suffers from short playing time – about ten minutes a disc as opposed to around thirty on the Philips – but it is less expensive and, what is more, whilst Philips and all the other system developers have been talking and promising, Teldec, as it is known, has been in the market-place working to establish the video disc as a credible idea.

Whether both the cassette and the disc can burst into the home market at almost the same time and survive, we shall soon know. Whether one standard for either will win through also remains to be seen. But the world's giants of electronics all recognise one thing. The latest major growth area in domestic electronics, the colour television receiver, is running out of steam. So many people now have colour sets there is little room left for expansion in that market. But the multi-million-dollar bonanza could start all over again if everyone who now has a colour set could also be persuaded to buy a video cassette, a video disc, or both, to go with it.

Digital Audio

Whilst there will always be development engineers and inventors seeking to bring us new concepts in home entertainment like surround sound and home video, there are also those anxious to improve on what we already have. It is in this field that one of the most important changes will come in the next decade.

Since the advent of sound recording, the signals, cut into the cylinders and discs, or laid down magnetically on coated tape, have been curves. These curves have been representations of the changes in sound recorded during the performance. They are called analogue recordings, and because they are actual representations of the sound they contain, they have needed top-quality mechanical equipment to record and reproduce them. A tape recorder with an unstable motor will produce a recording which has both unstable pitch and speed. A record deck must be kept free from dirt, because a speck of dust will feel to the stylus in the groove just like part of the curve it is following. The result: a perfectly

The frequency analyser which displays on your television set

One of the first generation of digital sound recorders

reproduced bang or click. And perhaps most serious of all to the listener pursuing sound perfection, even the most finely ground, evenly distributed tape coating produces a noise of its own – it is called tape hiss.

It would be wrong to give the impression that these difficulties are insurmountable to today's hi-fi enthusiast. Well-engineered record players and tape decks, ingenious dust-removing systems and quite brilliant noise-reduction units to reduce tape hiss go a long way to bringing satisfaction to the most fastidious listener. But hi-fi technology seems to have brought the overall ratio of music we want to hear, to noise and interruptions we do not, to an upper limit. To make a significant improvement now we need a rethink of the way sound is recorded.

The key to the new approach was to find a way in which sound curves could be changed into some form which could be recorded free of all the noise problems, and then changed again to be reproduced in their original glory. For several years now research engineers, with the BBC leading the field, have been investigating the possibilities of using a digital code, the language of the computer. It is by no means as difficult to understand as you might think, or as some computer engineers with their language of 'buzz' words would have you believe.

Looking at our analogue curve with its sweeps, its peaks and its troughs, it is easy to accept that it is a continuous line. If at regular intervals along that line we draw dots, the relative heights of those dots will change. The horizontal distance between them will be constant, but their relative height will depend on whether the curve is dipping or peaking or whatever. If we remove our original line and leave behind the dots, we are left with a pretty clear indication of the shape of the original curve.

All we are left with is the difficulty of transferring our pattern of dots on to tape or disc. We need a code. Because the horizontal distance between each dot is constant, all we need to identify each one is its height information. If the analogue curve was fed to a voltmeter and each time we took a height sample we recorded the voltage associated with that height, we would end up with a list of voltages, each one representing a dot position.

Our curve has become a list of numbers, and complicated numbers at that – no doubt several decimal points representing the minute voltages, and tiny changes between one sample and the next, only a micro-second or two further into the piece of music.

Computers handle complicated numbers using a binary code. A code with two possibilities: on or off. Written down, that can be 1 or 0, or in audio terms, a pulse or no pulse. There is a standard way of converting digits into binary code: 0 is 0 and 1 is 1, of course, but 2 in digits becomes 10 in binary, 3 in digits is 11 in binary and so on. So the binary code depends on a pulse for 1 and a missing pulse for 0. To achieve a steady stream of pulses and gaps that can be recognised, the code must be transmitted regularly and once that has been achieved it is not difficult to record the code on to tape or disc, or for that matter to transmit it along a cable or through the air as both the broadcasters and the telephone services are doing already.

The beauty of the system is that it can totally disregard any interference which gets in its way. Dust and tape hiss can cease to be serious enemies of sweet sound, because the code is lifted off the tape or disc during reproduction and the pulse pattern is used to generate the original sound. Providing the recording medium can deliver the pulse pattern – and machines have been doing much more complicated work satisfactorily for years – the digital to analogue reconverter takes care of the rest.

Initial work in the digital audio field indicates that the ratio of music we want to hear to noise we do not, will be improved substantially by digital techniques. Accurate representation of the original depends on the number of samples taken. A rule of thumb says twice as many samples must be taken every second as the maximum frequency in the recording. The BBC radio bandwidth is 15 kilohertz, so to leave a safe margin the BBC currently samples 32,000 times a second.

All the control pulses are added in to the original code, to keep the whole system in synchrony, which means an awful lot of pulses are being transmitted. Perhaps as many as a million a second for two stereo channels.

Today the only equipment capable of handling such bandwidths is video equipment. The current VCR has a bandwidth of 3000 kilohertz, or 3 megahertz. Philips, which has been about to launch its videodisc for at least the last four years, is keen to exploit its possibilities in the audio field as well. This has a $5\frac{1}{2}$ megahertz bandwidth – the spare frequency capacity could be used to extend the playing time of the disc, or to include extra information for more than two stereo channels, always assuming surround sound of that type is still in vogue. At least one Japanese manufacturer has an add-on sound unit for its home video equipment, and all the signs are that the pressures to turn from analogue to digital will be put on all of us in the not too distant future.

Incidentally, at the time of writing this chapter the Audio Engineering Society has set up a committee representing manufacturers interested in digital recording techniques. Though this committee will, initially at least, concern itself with the problems of professional equipment, the aim clearly is to establish standards throughout the world. Hopefully, this will mean that digital audio will avoid the difficulties which beset quadraphonic sound with its many systems and standards. Compatibility between different manufacturers' equipment and recordings seems little to ask, but may prove difficult to achieve.

I hold firmly to the belief that the home-sound business is like the Emperor's new clothes. There is a level of improvement in sound reproduction above which we are incapable of detecting any improvement because our hearing is not sensitive enough. And yet enthusiasts are prepared to be persuaded to try new techniques and systems which will raise their level of enjoyment and appreciation just that little bit more. What we can all hope for is that in seeking to increase the overall range of sounds we can record and reproduce, new technology will also make improvements in the range that all of us are capable of hearing.

If you are intrigued by what differences digital techniques can make you might be able to sneak a preview of the next decade if you are lucky enough to come across a series of Japanese discs cut from digitally-recorded tapes. Denon, a Tokyo-based record label, has used a modified professional video recorder to make what amounts to some demonstration records. The discs, of course, are analogue as usual, but the master tapes were recorded digitally using a sampling rate higher than the BBC's.

The catalogue extends from Vivaldi through Bach and Telemann to a group called The Grand Spaceship Orchestra. The disc I have heard is of distinctly middle of the road material, but the clarity and fine treble response on *Ebb Tide* and *From Russia With Love*, could not be denied. I feel that pulse code modulation – electronics jargon for the technique I have been describing – will take some getting used to, because the exceedingly clear sound is also an unrealistic sound to the ears of anyone used to conventional recordings. Tests in America using equipment developed jointly by 3M and the BBC produced startling results. The sounds of instruments rubbing against the musicians' clothing, the noise of fingers on the keys, were so faithfully laid down by the digital recording that new microphone placings quickly had to be worked out.

Any system that comes close to total honesty may be difficult to live with, but the beauty of hi-fi is that we each hear things through a different set of ears. We are searching for what sounds right to us, not necessarily for what the development teams can prove sounds good on paper.

One eventual development of digital recording may be the end of discs and tapes as we know them. It would be quite possible to store the coded information to generate quite long pieces of music on plastic cards no bigger than our current credit cards. The recordings could be holographic or magnetic – and the scanning system for picking them up could have very few moving parts. One result, apart from whatever improvements there are in the actual sound, would be greatly increased storage capacity in the conventional record cupboard. That may be a more significant advance than would at first sight appear.

Seeing the Sound

To return to making the most of what we already have. One Swedish engineer has made it possible for us all to cut through the sales talk and technical jargon and to see for ourselves one area where our present hi-fi set-up is deficient.

Ideally any hi-fi system should produce an even reproduction of all sounds, high and low, with perhaps a slight reduction in treble frequencies to avoid the recordings and broadcasts sounding too shrill. This is what experts call producing an 'even frequency response'. But believing your own ears can be difficult, so an imaginative Scandinavian has produced a unit which can display a frequency response pattern on our television screens, indicating how our sound systems operate in our own homes. And the first bit of that sentence is important because the best equipment in a shop or a friend's home may sound quite different in changed surroundings.

A noise generator, which produces a rushing sound made up of all the frequencies a sound system is ever likely to have to reproduce, is connected to the amplifier. A microphone in the listening area picks up the noise and analyses how the amplifier and the speakers are handling each frequency band. Then it uses one of the spare channels on the 625-line VHF television band – there are plenty of them – to reproduce a series of vertical columns. Each column represents one audio frequency band.

Ideally, each column should be the same height, indicating even frequency response, with that gentle rolling-off towards the higher frequencies. Any column sticking out above the rest indicates that one part of the sound spectrum is being emphasised; any column shorter than the others means that area is being neglected.

One way to rectify matters is to adjust the tone controls of the system, but there could be another reason for exaggeration or neglect of one area of the frequency spectrum. Sometimes the sound signals do not travel a direct path between loud speaker and listener. They are reflected off walls and ceiling or lost in curtains and carpets, and whereas few enthusiasts would go to the lengths of a total alteration to the domestic décor, it is interesting to see what differences slight changes can make.

Moving the loudspeakers is the most likely source of noticeable change; away from the corners perhaps, or higher off the floor. All the while the display on the television set indicates whether an improvement has been made in that flat response line.

Only the most ardent sound fan is going to purchase such a device, but it could be useful to those selling hi-fi equipment and servicing, because it will be possible to demonstrate to the user that an improvement has been made or that things are as good as they ever will be. Seeing with our eyes as well as hearing with our ears will be a valuable weapon in the fight against the Emperor's new clothes.

'Lasers Entertain'

It takes a great deal of time for most of us to come to terms with major new developments. The computer industry has had an uphill struggle persuading those outside the trade that it can be an invaluable tool to all of us, and the same can certainly be said of the laser.

Dreamt about in the fifties, made to work in the sixties, the laser quickly established a position in most people's minds as an instrument of destruction. And not without some foundation. Light from a laser can be a narrow beam, straighter than any ruler; depending on the energy put into the laser, the beam can be highly destructive.

Now, in the latter half of the seventies, the laser is becoming better understood, and a generation of artists is using its unique properties to develop new creative forms. It is even possible to do what only a year or two ago would have been unthinkable to many: bring a laser into our homes and use it for entertainment.

But first let us take this opportunity of reminding ourselves what a laser is and why it has these properties. Light amplification by stimulated emission of radiation – the description from which 'laser' comes – is not as fearsome as it sounds.

Inside a glass tube a substance, which is often a gas or mixture of gases, is excited with electrical energy. This stimulation makes the atoms of the substance fluoresce; they glow, producing light, the colour of which varies from substance to

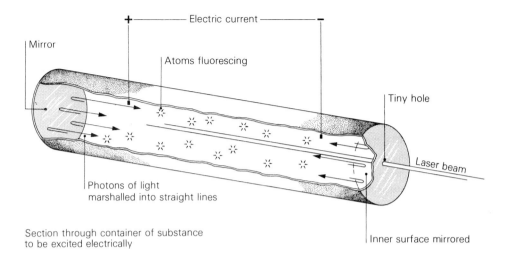

Section through container of substance
to be excited electrically

substance. At both ends of the glass tube are mirrors, so the light produced by the stimulated atoms is reflected backwards and forwards inside the tube and as the light from one stimulated atom meets the light from another the overall light output is amplified.

This activity inside the tube gradually arranges the photons of light into straight lines. The mirror at one end of the tube allows some of those photons to escape to become the beam of the laser. That beam is made up of light of just one wavelength. This is monochromatic light with each photon of light energy following the same wave pattern down the beam.

Anyone who watches *Tomorrow's World* regularly will be familiar with the pencil line of intense energy that makes up a laser beam, but even television can do little justice to the bright image the beam produces when it strikes an opaque surface. It is this brilliant image which attracted the attention of artists. Could the clear, sharp, almost hypnotic pinpoint laser image become an art form?

The first laser image for entertainment has already delighted theatregoers sufficiently to suggest that soon we may all find 'going to the lasers' as much a part of our leisure time as going to conventional shows is now. The Americans have led the development by producing equipment which enables the laser beam to be scanned over wide areas very quickly so giving the audience not one tiny dot, but a series of dots painting a laser picture.

Just like the cinematographers before them, the laser artists depend on our natural persistence of vision to achieve their effect. The dot of the laser beam produces such a sharp stimulation in our visual nerves that our eyes can still see that dot long after the beam which produced it has moved on to make other dots close by. The result to the viewer is a line of bright laser image, which can be moved and developed to produce words, shapes and complete pictures.

When one fast-moving laser dot is allied to several others, perhaps in different colours, the artist has a real palette of light with which to paint his new pictures. A mixture of krypton and argon gases in the glass tube from which all this starts, will produce a very bright, white beam. A prism will break the white light into its component colours. Such colours will be pure and coherent and these become the palette for the laser painter.

Now a series of mirrors is needed to direct the colour beams first to the mechanical and optical devices which will scan and control them, and then on to the display surface itself. Crystal filters alter the intensity of the beams. The more dense the filter, the more energy is absorbed from the beam, so filters have been developed with a variable density. Straightforward gates can be flicked in and out of the path of the laser, switching it on and off, or flashing it very quickly. Interference and effect filters are then dropped into its path, duplicating the beam perhaps or giving it a halo, and finally the beam reaches the scanner itself.

The fast movement of the beam can be achieved by two mirrors mounted at $45°$ to one another. These are movable, powered by galvanometers which rock from side to side. The beam strikes one mirror, is reflected on to the second and off towards the display surface. By operating the galvanometers in harmony with each other the beam can be deflected very quickly over an exceedingly large area.

The controls for such a set of a mechanical and optical devices have to be exceedingly fast-moving as well, but in the age of the computer such speeds are well within our capability. So the laser artist, or laserist as he is now called, produces a programme of effects and images, shapes and pattern, which is then digitised and put on to magnetic tape.

That tape becomes the set of instructions for the laser projector; each coded control signal becomes a closing gate, a change of filter or perhaps a variation in the movement of the galvanometers. A totally automatic laser show now only needs a music or speech recording running in synchrony with the digital control track to be complete.

But there does seem to be a growing, and I feel most welcome, trend in the laser business. That is to de-automate the entertainment and put some of the control at least into the hands of the laserist himself in the theatre or auditorium. Give him or her the same sort of projector and control system but provide a much more basic digital master track. Then the frills and final interpretations can be put on to the show by a performer sitting at a control desk which is also linked to the laser projector's filters and galvanometers.

The temptation to put the whole programme into the can, so to speak, is a great one because it means that one successful laser artist can have shows running simultaneously in Munich, London and Los Angeles. But the move towards personalising the laser show and doing each performance live on the night with the artist present with the musicians to produce the sounds, will appeal to anyone who works on *Tomorrow's World*. It is a programme that has long resisted the technology of recordings and still transmits live. That way everyone shares the genuine experiments as they happen, not as some computer or recording machine could make them appear to happen.

Christmas 1978 saw the laser in London's Oxford Street delighting the crowds of shoppers. *Tomorrow's World* reported on the struggle the traders' association had to convince the local council that the display was a safe one, but when approval was given the sight was a genuine crowd puller. No music or spectacular patterns, simply straight lines of red and green light slicing through the darkness just above the tops of the double-decker buses.

After Twelfth Night the Oxford Street display had to go but as a permanent feature the laser show might serve a very valuable purpose in bringing new life to the many theatres and cinemas hard pressed to survive on conventional entertainment. Already, however, a home version of the laser show seems inevitable. This involves using an exceedingly low-power unit, almost certainly less than a milliwatt, which might be fitted with two revolving optical surfaces operating like the galvanometer mirrors in the professional projector. Effect slides and filters could be inserted between laser projector and the mirrors, and the controls could be linked either directly to a sound frequency splitter, not unlike those used in discothèque light shows today, or to a series of switches for the home enthusiast to operate by hand. With the introduction of home lasers as well as home video recorders to our firesides it does seem that the temptation to be unadventurous and stay in during our leisure hours is going to be even stronger in the eighties than it was in the seventies.

Nights on the Town

But it would be wrong to give the impression that I see leisure in the future as a totally stay-at-home activity. There will still be experiences to enjoy which can only be found away from the family hearth. The cinema business will not give up without a fight. The most recent box office smash hits, including *Star Wars* and *Close Encounters of the Third Kind*, have all had almost as much attention to their sound tracks as they have to their pictures.

For instance, *Star Wars* was shown in London with a sound track consisting of six channels of independent signals, each channel treated with the now famous noise reduction system of Dr Ray Dolby. If this is new to you let me say it means, briefly, that in recording, the quiet passages are artificially amplified, and subsequently attenuated in reproduction. The over-all sound is unchanged, but unwanted noise added in the recording is greatly reduced.

We have already dealt with the difficulties of getting four channels into your own home. Six must be worth an evening out! Cinemas in other big towns may only be showing the same films in noise-reduced stereo, and I am afraid there will be theatres showing just the ordinary monophonic version.

But it is an indication of things to come in the cinema, and if you are interested in getting real value for money it might be worth checking which system your cinema uses.

I close with something you can certainly not experience at home. Not that it is anything new, but it is a leisure activity which is enjoying a tremendous revival in the United States, and which recent technology has made even more exciting. I first experienced it in Santa Clara county at the southern end of San Francisco Bay. At each end of the Tidal Wave are two towers, approximately 140 feet high. A steel track connects the tops of the two towers which stand some 400 feet apart. In that intervening distance the track drops to ground level and before rising to the top of the second tower does a complete loop through 360° – the highest point of the loop being some 50 feet above the ground.

A small train of cars sits on the track, stationary, at ground level. Behind it a steel weight is wedged. On the end of the weight is a hawser which is wrapped around a winch just underneath the loop; the winch is also connected by hawser to a 45-ton weight sitting at the top of one of those towers.

After the train loads with riders the brakes are released and the weight plummets to the ground spinning the winch as it goes. This in turn pulls on the smaller weight behind the train. The result – the train accelerates exceedingly quickly pushing all the riders back into their seats. It is doing around 55 mph less than five seconds after starting, it climbs into the loop, goes over the top, rushes down the other side and up to the top of the 140-foot tower. For a second at the top it pauses, then gravity wins and back it comes, its riders now facing away from the loop, and through it goes once more. On it goes through the loading point to the top of the other tower, again a change in direction and back into the loading station, where it stops thanks to the system's pneumatic brakes.

In all, ninety seconds of stomach-turning, or as the Americans call it, 'knuckle-whitening', thrills. And then the patrons leave the ride to join the back of the queue to ride again.

It turns out that this latest attraction of the American Theme Park, as their modern fairgrounds are called, is actually an up-to-date version of a wooden ride found in Coney Island before the turn of this century. Then it did not catch on. Just another example of a good idea before its time.

Incidentally, if the idea appeals, I believe that there might be a 360° loop built here in the United Kingdom in the not too distant future. *There* is a sobering thought!

Michael Blakstad

Whatever Happened To . . . ?

'Whatever happens to all those inventions you demonstrate on *Tomorrow's World*? How many of your predictions come true?' They are good questions, the most common we are asked, and the answer in itself amounts to a statement of the programme's intent.

It would be easy to think of a number (somewhere around 5 per cent is about right) and trot it out, but to do that would be to dodge the main issue. If the 'strike rate' were the most important criterion against which the programme's success was judged, then we could raise it at once by the very simple device of selecting items which are about to appear in the shops, the products of intensive development work and market research, backed by the most professional companies in the business. Then we could rename *Tomorrow's World* and call it *Shop Talk* or *Jim's Inn*, and most of the people working on the programme could put their feet up, leaving their jobs to be done by the eager advertising men whose products we were gracious enough to promote.

But that is not the philosophy of *Tomorrow's World*. The further we can travel away from the market-place, back down the road towards the laboratory, the drawing-board, to the men and women who are presently dreaming up the ideas which will one day, perhaps, find their way into our daily lives, then the happier we are. Give us a prototype, a breadboard model, a test-tube experiment or simply a problem which still needs solving and you've achieved the top rung of our particular set of criteria.

The corollary of this is, of course, that the nearer we get to our goal, the less chance there is that the items we demonstrate will subsequently succeed. There is many a barrier between the research bench and the high street, and in today's competitive world even the best ideas are all too often starved of funds or prove too expensive for the market to bear. The further *Tomorrow's World* retreats from the final, double-cellophane-wrapped produce on the supermarket shelves the less chance there is that our predictions will come true.

Subjects are often chosen precisely because we believe something is going wrong, or is likely to fail. *Tomorrow's World* was born in 1964 when Britain's first Minister of Technology was persuading his Prime Minister that the nation was to be supercharged by 'the white heat of technology'. Harold Wilson was the Prime Minister, and Tony Benn the Minister of Technology. The white heat of technology then began to turn red and finally dull black. Grandiose schemes like Concorde and Britain's multiple nuclear power programme were proving to be far too expensive for an economy like ours to sustain, and the public was learning that scientists and technologists, when left to their own devices, could make terrible mistakes and spend huge sums of money doing so. Inevitably the programme's philosophy has changed. *Tomorrow's World* still reports good news where it is to be found – and happily the good still far outweighs the bad in our output – but there has been an increasing emphasis since the last of these books on reports which demonstrate that shaping the future is a very precarious way to earn a living.

The preceding chapters have been built on the team's experience of the struggles and ambitions of today's scientists and engineers; we have reported 'the state of the art' and based our predictions on our knowledge of other people's work. In this final chapter, we move one step closer to the inventors, entre-preneurs and innovators on whom we have always relied for the programme's bread and butter, the steady diet of over 2500 items reported in more than 500 editions of *Tomorrow's World*. 'Whatever happened to such-and-such' is nearly always the story of human as well as technical triumph or disappointment. Here we report on five projects, four British and one American, and, most important, on the people and the dreams on which they were based.

Whatever happened to . . . 360 Degree Scissors?

The Saxons invented them, and the ancient Chinese developed the shape we know today, so you would hardly expect to look to scissors for lessons in the harsh school of innovation. But the Saxons and the Chinese did leave one problem unsolved, and that is – what happens when scissors fall into the wrong hands? It is worse for left-handed people who cannot buy scissors made specially for them, but even right-handed people can share the frustration when they try using their scissors in the left hand to cut, for instance, the nails of their right hand. The clenching action which pulls the blades together if you have them in the correct hand acts adversely and pulls the blades apart, away from the pivot, if you use them in the wrong hand. To test this, simply take some ordinary scissors and put them in the hand you do not normally use and then try to cut a thin sheet of paper: because your clenching action is now pulling the blades *away* from each other, the paper will very likely slide between the slack scissors.

A Devonshire designer who had been mulling over this problem worked out what he thought was the perfect solution; he even allowed himself the hope that it might make him rich. His idea was to allow the scissors to swing beyond their normal semicircle and carry on round for a full 360°. Both blades were designed to have a cutting edge on two sides, and the handles were carefully shaped to give a perfect configuration for right-handed cutting at one meeting point, and for left-handed cutting at the other.

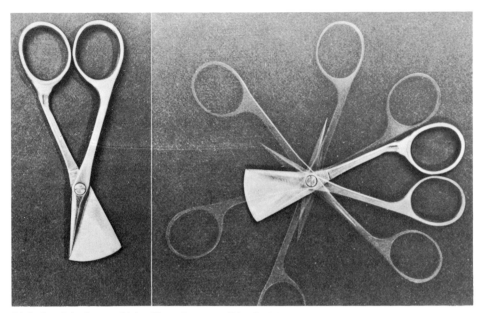

Right-handed scissors which will cut just as well in the left hand – or vice versa

Richard Hawkins had spent eight years as an insurance broker at Lloyd's, saving up for a metal lathe and what he describes as a dream workshop. 'My heart,' he says, 'was more in the workshop than in insurance and the break came when I met my wife: she had worked in Devon and we decided to move down there and try to make a living in the country.' Hawkins gave up insurance and took a job at an electrician's shop in Kingsbridge, while his wife took up wood-turning. After a while, he began to manufacture solar panels and later set up his own tiny company.

He worked out the best shape and method of operation for his 360° scissors by means of cardboard prototypes; the final design evolved slowly and when he reached a design which satisfied him, he simply put the prototype on one side – he had been more interested in solving the problem than in making a working model in metal. It was his wife who one day, when bored by the cardboard scissors which were just littering the workshop, announced that she had made an appointment with a patent agent in Plymouth; Hawkins rapidly machined a metal prototype.

Patents, though they were not to know it then, were to be the biggest single problem facing them. Any new idea, and especially one as simple and ingenious as this one, is likely to have occurred to someone else at some stage, and that someone might well have filed a patent. To give the inventor a brief period during which he can search the records without other people copying his idea, pro-visional patents give one year's respite, a 'truce' during which no one can borrow the design. It was in January 1977 that Hawkins and the patent agent secured a provisional patent specification on the 360° scissors, so he had a year in which to talk with possible clients, discuss manufacture, show his design on *Tomorrow's World*, and modify his idea so that any changes could be incorporated into the complete patent specification a year later.

He faced a year of almost constant commuting from Devon to both Sheffield and London, the cities in which his scissors were likely to be, respectively, manufactured and sold. Hawkins found Sheffield an eye-opener. The mixture of huge and small steel works must in itself have been a surprise to the Southerner, but the biggest shock was the nature of the Sheffield scissors industry. True to the city's tradition of master cutlery, the whole process was 'made by hand'. The first step was to have a 'blank' made – a master matrix – and there was only one firm which makes blanks; the price was therefore high. The next stages, the grinding, polishing and assembly of scissors, were performed by a number of firms, and every process was performed by a different set of hands; all 'hot-forged' scissors in Sheffield are made today just as they have been made for centuries; the final stage of setting, the crucial step during which the blades are curved to make sure they cut, is a highly skilled job performed by one man with a hammer.

Several visits to Sheffield, the manufacture of that vital blank, and a number of trial assemblies assured Hawkins that manufacture was at least possible, although likely to be expensive. The other job was to establish whether or not there would be a market for the new scissors. After visiting virtually every department store in town he won the promise from every buyer but one that they would be very likely to retail the scissors when they became available. It was as much as one man could expect to achieve on his own: his income had dropped to almost nothing, he had spent almost £5000 of his own (and his bank's) money.

In the meantime, Hawkins had been pursuing one other line of enquiry. Wilkinson Sword, the British steel goods specialists, had long since moved on from manually assembling scissors and had installed machine production lines which churned out huge quantities for a mass market. If he could interest them in his 360° scissors, then – although he would lose control of the project – at least he would be assured of bulk manufacture and marketing. Wilkinson's were very impressed by Richard Hawkins' design, so much so that he thought they were going to write him a cheque there and then, but they had one proviso. Pointing out that it would cost them a quarter of a million pounds to set up the machines to produce his design, Wilkinson's insisted on conducting their own search for existing patents. They soon discovered that the files did indeed hold a similar idea.

The date on the previous patent was 1926, which Hawkins and his Plymouth agent believed should cause no problem; British law only requires the applicant to search for fifty years back into the archives, so for the United Kingdom the old patent had effectively lapsed. Wilkinson's interests, however, stretched far beyond the domestic market; manufacture on the scale they planned depended on control of the world market, and American law requires a hundred-year search. They were at perfect liberty to manufacture or to sell in the United States; the old patent had not been taken up and was therefore reckoned to be dormant. What they could not do, however, was to prevent other manufacturers making absolutely identical scissors and selling them there. Wilkinson's at this point withdrew from the scene.

This left the inventor with a king-sized decision to take. A patent protects not the origin of manufacture but the market. A British patent therefore protects only the British market, leaving the rest of the world open to any foreign competitor who comes across Hawkins' design and decides to 'borrow' the design. The Germans and the Japanese, accompanied by a healthy number of Americans, are known to be avid students of British patent applications, knowing that as often as not the British inventor will be unable to afford the £25,000 it costs to patent a product across the globe.

So, unprotected by patents except here in Britain, Hawkins had to decide whether it was worth embarking on his own upon the manufacture of his scissors. The smallest number he could commission from the cutlers of Sheffield was 6000. On top of the professional fees due to accountants, patent agents, solicitors and the others who are necessary to launch a new enterprise, the cost of entering the scissors business would rise well above £10,000, which went way beyond any securities Hawkins could offer his bank for the overdraft. What was worse, he discovered that the department stores were not going to be very generous in their terms. The buyers had indicated that they were prepared to agree to the price he named but now he discovered that they would not pay him until twenty-one days after delivery and that they expected a 'settlement discount' of 5 per cent. As Hawkins puts it, 'I am expected to lend them the money for twenty-one days, and they then deduct 5 per cent for the privilege!' To make matters worse, they demanded a banker's reference, and Hawkins could not be sure that the bank would be backing him.

Despite all these difficulties, in March 1978, Richard Hawkins took the plunge and decided to go ahead with his first order of 6000, hoping to get them into the shops by that September. If they sold by Christmas, he would plough the profits back into the manufacture of another 10,000, and hope that from there the sale of the scissors would expand to the next critical point where he would have to decide whether or not, unprotected by patents, he should try for the overseas market. He hired a new accountant, and the accountant has succeeded in persuading Hawkins' bank manager that the gamble is worth backing. Hawkins has been lucky in one respect – scissors are small enough for one man to cope with, and success or failure depends very largely on his own ability to mastermind the manufacture and to sell the product. Our second pioneer has not been quite so lucky.

Whatever happened to . . . ALZA?

If the path for the British innovator is strewn with obstacles, it may be reassuring to know that there are problems on the other side of the Atlantic as well, even in the heartland of American new technology, California's Silicon Valley.

Next door to the ever-prospering Hewlett Packard is the equally resplendent headquarters of the Alza Corporation, a business venture built on one scientist's dream that he could direct medicine more efficiently to the diseased or injured part of the patient's body.

If there is a flash fire in a chip pan, it does not make a lot of sense to hose every room in the house, ruining the furniture and the fittings, where one well-aimed extinguisher could have done the job. Yet, reckoned Alex Zaffaroni, that is precisely what we do to our bodies every time we take a drug: instead of delivering one well-aimed dose to the precise point where the trouble really lies, the conventional pill, dose, or inoculation saturates the entire system in the hope that a small portion of the medicine will reach its target.

It was upon the concept of 'targeted drugs' that Zaffaroni built Alza (Alex Zaffaroni). He is Uruguayan by birth, and after studying medicine at Montevideo University, came to the University of Rochester to win his doctorate in biochemistry. He was brought to Palo Alto by the then modest-sized Syntex organisation. As Syntex' sales rocketed up to $75 millions in the late 1960s, Zaffaroni was developing both a healthy appetite for the American way of business, and his entirely novel philosophy of drug delivery.

Zaffaroni looks at it like this: the range of drugs already researched and developed is now so wide that only the biggest and boldest corporation can undertake the hugely expensive undertaking both of discovering new drugs and of satisfying the Food and Drug Administration regulations in America, or equivalent watchdog bodies in other countries. Instead of attempting to add to the already bulging pharmacopoeia, his idea was to make his fortune by devising new methods of delivering medicines. In 1968, he left Syntex and moved one block away to set up his own organisation.

The first product he developed is the Ocusert which could, believes Zaffaroni, become the most potent weapon yet in man's battle against the widespread eye disease, trachoma. The idea of the Ocusert is simple: instead of swamping the diseased eye with a giant dose of powerful medicine with a number of eyedrops given, say, in the morning and at night, with nothing in between, Zaffaroni devised a minute 'sandwich' of transparent plastic which sits on the eye, gently releasing the dose in precise measured quantities regularly throughout the day and night. The device is now in limited manufacture, and is being used in some Western countries to combat glaucoma, another serious eye disease, but – and it's a big but – the device is far too expensive for the Third World countries where trachoma is prevalent, where expert medical attention is all too often lacking, and where the Ocusert, Alza believes, is just what is needed as it can continue to release its medicine for a week at a time, without supervision of any kind.

Zaffaroni describes the situation as the perfect Catch 22: 'With any product involving plastics,' he says, 'you cannot bring the price down until you are manufacturing in bulk, but you can't manufacture in bulk until you have a huge demand, and you don't have a huge demand until you bring the price down!' At present, Alza is experimenting with a simpler design of the Ocusert, using one layer of plastic instead of three, in an attempt to bring production costs down. Until then, it will remain the prerogative of the richer patients.

Another 'targeted' delivery system is the Progestasert, and this one has hit even more serious snags than the eye sandwich. The idea behind this device is that women would suffer fewer side-effects from taking the contraceptive pill if the progesterone were delivered directly to the uterus and not released to other parts of the body. The Progestasert looks like a standard intra-uterine device; when in position, Progestasert releases its hormones in a regulated dose for an entire year. The quantity of progesterone which enters the body through this system is minute – the equivalent of only three normal pills in one entire year – and the evidence is that they do prevent side-effects. The FDA duly passed the Progestasert, and all systems seemed ready to go for the device, until the devastating news in December 1977 that it had one major shortcoming. Like all contraceptives, the Progestasert is not 100 per cent safe, but this was not the bombshell. What put Alza back on its heels was British evidence that, among the few pregnancies which could be expected amongst women using the device, the chance was high that these would be ectopic – that is, the embryo would be formed outside the womb. It is a perfect illustration of the hazard of launching new products in the life sciences field. As Zaffaroni says, once you establish a product, you're clear of the field; but it's the most difficult branch of industry in which to get a new concept off the ground.

Zaffaroni seems undaunted. He is working at present on perhaps the most ingenious of all his drug delivery systems, a tiny 'pump in a pill' which, when swallowed, travels through the digestive tract, pumping out its medicine in precise, regular, measured doses for twenty-four hours or so until the pill passes out the other end. The pill consists of an outer skin of semi-permeable membrane: it permits the body fluids to enter the pill but not to escape. The drug is packed inside the skin in soluble form, so when the liquid is attracted into the pill by osmosis the drug dissolves and the volume inside the pill increases, setting up an 'osmotic pump' action. It wants to get out and escape is possible through a tiny hole etched into the skin of the pill by a laser. The size of the hole dictates the speed of emission of the drug, just as the time of release, according to the organ needing the medicine, is controlled by the speed with which the drug compound is designed to dissolve. Thus, the pill travels right through the body emitting its tiny trail of medicine. This is better, believes Alza, than the conventional pill which sits on the lining of the stomach and delivers a large dose, virtually all at once, and then the stomach is left for a few hours until the next pill arrives and delivers another hefty dose.

But it has all cost money. About 500 top-grade scientists and laboratory technicians work for Alza, occupying some of the most beautiful factory and office buildings in the not unprosperous Silicon Valley. By December 1977 Zaffaroni had managed to raise no less than $100 million of venture capital to support his projects, without any real prospect of a successful market product for a number of years to come. In December, Ciba Geigy moved in with a further $35 million with what, to all intents and purposes, represented the takeover of Alza. Zaffaroni and his scientists work on, with new projects bubbling up in one of the most creative scientific environments anywhere in the commercial world, but the story of Alza's first ten years illustrates only too graphically the economic problems facing the innovative scientist or inventor when he tangles with the unknown and the unproven.

Whatever happened to . . . Tony Dawson's motorcycle frame?

Whatever the difficulties which beset Alza, the story does illustrate one crucial difference between the British and the American scene. For reasons as varied as the attitude of financiers, the diverse taxation systems, the availability of private wealth, and a host of others, the stark fact is that venture capital is available to the American innovator, whereas in Britain it is much, much harder to find. The prosperity of Silicon Valley in California, of Cambridge's famous Route 128 in Massachusetts, of fresh innovation centres littered across the 'Sunshine Belt' and elsewhere in the USA, bears evidence to the willingness of financiers in America to gamble on new technologies and of the American public to buy new products.

The much more sombre British scene is well illustrated by the story of Tony Dawson, an engineer who lives in Mallin Bridge, Sheffield. When *Tomorrow's World* came across him in 1975, Tony Dawson was on the dole, and was using his unwanted free time to think hard about the way in which motorcycle frames are designed. A lot of time and skill goes into the welding of the many joints and bends which go to form the complicated tubular structure of today's frame, and Tony Dawson reckoned that the cost of making every frame by hand could be reduced by the simple device of stamping two flat frames directly from sheets of aluminium and bolting them together. The same frame could be used for almost any make of bike. He put a great deal of work into the metallurgy and into stress testing and came to the conclusion that his bike frame would not only be a lot cheaper to

Designer Tony Dawson assembles his revolutionary motorcycle frame in the Tomorrow's World *studio*

produce, but was easily as strong and as safe as the more cumbersome conventional models.

To prove his theories, Dawson himself built a racing prototype around his frame, fitted with a 125 cc, twin cylinder engine capable of developing 25 brake horsepower. It was completed in 1974, and was immediately put to the test by Steve Machin, twice British racing champion and a close friend of Dawson, who believed very firmly in the concept of the new frame. The first test run was on 24 July 1974. Everything worked perfectly until, satisfied with the bike's performance, Machin insisted that the inventor himself should have a go. As Dawson struggled to start the machine, Machin roared off on another make of motorcycle.

At 95 mph his gearbox seized, the back wheel locked, and he skidded 130 yards before crashing. He died shortly afterwards, a great loss to British motorcycle racing and, to Tony Dawson, both a personal tragedy and a professional disaster. He simply gave up the bike frame for a period of six months. 'I'd lost a very, very close friend,' he said, 'and, with Steve being really the driving-force behind the project, it left a lack of incentive. Then, after a while, I decided it had to go ahead, Steve would have wanted it to go ahead.'

Dawson's frame was shown on *Tomorrow's World* in October 1975, and before we came off the air there had been several phone calls from people interested in backing the venture. One of these was from Guy Harmsworth, a relation of the press baron and a man so anxious to play a part in reviving Britain's moribund motorcycle industry that he offered to sell his house and raise every penny he could to help the project along. In all, he has spent £18,000, mainly on agents, patents, legal fees and on getting advice as to how to raise the rest of the money needed to get the frame into production.

Their first action was to form a new company to build their first production model, the Manxman, a 250 cc road machine with emphasis on easy maintenance, economy and, of course, safety. They planned at first to go into small-scale production in the Isle of Man because it offered attractive development grants and fairly easy-going regulations for the entrepreneur setting up a new company. Unfortunately, the two men discovered that the cost of transporting parts and equipment from some thirty-five suppliers on the mainland completely offset any advantages the island had to offer and would push the price of the Manxman above that of its Japanese rivals.

If the bike was to succeed it had to be produced in England, so Dawson and Harmsworth set off on the trail of venture capital. They found it surprisingly hard to find – so much so that they even had to resort to a consultant who could tell them where to begin and which finance houses were likely to consider providing start-up money to entrepreneurs with no track record in running a business. The harsh answer, they discovered, is that there are none. The nearest they got was a merchant bank which offered to come in a bit later on if someone else would put up the seed money, but that was all.

By October 1977, Harmsworth's money had disappeared and Dawson was back on the dole. He's now working on a motorcycle wheel made, like the frame, from two aluminium plates held together and tensioned by screws around the rim. It looks as good as the latest cast wheels while being easily as safe as the old wire spoke wheels, and Tony Dawson still dreams of the new all-British motorcycle of which the frame and the wheel will both be part. When *Tomorrow's World* once again reported on the plight of Dawson, Harmsworth and the frame, there was another flurry of interest from about forty companies, this time including a call from GEC, the new owners of the Meriden motorcycle factory and Britain's last large-scale producer of bikes. GEC has a hard-edged attitude to spending money; if there's going to be no firm chance of profit, it will not invest, and at the time of writing it appears to be more interested in shopping trips to California than in buying British companies. In May 1978, Tony Dawson was offered free factory space by the city of Sheffield, and he set about equipping it for production, but as he wryly put it, a factory is not much good if you haven't got the money to operate it! It will be interesting to see the outcome of this particular venture, an all too typical example of a lone inventor struggling to raise that most scarce of British resources, start-up money for new technologies.

Whatever happened to . . . Concorde 1?

On 15 January 1976, *Tomorrow's World* went supersonic. The programme took an 'awayday' to Casablanca in Concorde, which was then on the point of embarking on its first commercial service. It was a week in which Britain felt very sanguine about the Anglo-French airliner and, judging by the correspondence which followed the transmission, the Great British Public did not want to hear any but the good news about the twenty-year-old project into which the tax-payers of Britain and France had poured over £1000 million.

And there was good news to be told. Raymond Baxter reported on the technical achievement which had made the flight possible. The thrust of the Olympus engines, first on take-off and then when the stage two re-heat takes Concorde through the sound barrier, gives passengers the feeling of riding in a highly-tuned sports car when the driver puts his foot down. At Mach 1 the engineers had to cope with the fact that the air-flow over the wings alters and the whole centre of lift in the aircraft shifts, which meant that twelve tons of fuel in those graceful delta-shaped wings had to be pumped from forward to aft, at the rate of a tonne a minute, in order to compensate for the change of balance. At Mach 2 and 52,000 feet the air outside the aircraft is minus 61°C, but the temperature of the skin of Concorde has been warmed by the friction of air rushing past to a temperature of 108°C. Passengers can feel it by touching the inside of the heavily-insulated windows, and this rise in temperature actually stretches the aeroplane by a foot ($30\frac{1}{2}$ cm). To prevent the craft being torn to pieces by this expansion the designers have built expansion joints at one-metre intervals, just as bridge designers do. There is no doubt that Concorde is an extraordinarily fine technological achievement, and the programme duly reported that fact.

It was the second half of the programme which upset a large number of viewers, and particularly those connected with the Concorde project.

'Any other nation,' wrote a Gloucestershire MP in a letter to the BBC, 'would demand support from a broadcasting organisation which relies on public funds.' And he went on, 'I know that such judgments about programmes are bound to

The last Concorde has now been built, and the production lines closed down

affect the views of my constituents and myself on the review of the desirable arrangements for broadcasting in the future.

'History will show in due course,' he wrote, 'when we see how the Concorde and its supersonic competitors and successors (which there will be) fit into the world aviation market.'

Well, history does not take place over a mere two years, but the time seems ripe for an interim report on Concorde's performance to compare with *Tomorrow's World* predictions.

At the time, matching the heady mood of the nation, aviation correspondents of both press and television were content to rely on the performance and cost figures supplied by the publicity officers: noise measurements which proved that, with skilled pilots, she would not be any louder than older conventional jets like the 707 and Britain's Trident; range-payload figures which told of 120 passengers jetting over the most profitable routes and claiming that a new generation of high-speed executives was queueing up to fly to Bahrain and subsequently to New York and Washington.

Tomorrow's World suggested that some of this was acceptable, but some was misleading. As to noise, there was little point in dwelling on the subject: William Woollard reminded viewers that the noise was inherent in the Olympus engine; it was the best choice available at the time, but the decision had been taken too soon to take advantage of the new generation of quieter by-pass jets which eliminate a lot of the 'tearing' sound at the tailpipe of the engine. The worst thing about Concorde's noise is the 'footprint' – the size of the ground area which is affected by the sound of the engines on takeoff. Concorde's noise footprint is reckoned to be six-and-a-half times larger than the equally noisy Boeing 707, but the sad fact is that the 707s are on the way out, making way for newer, quieter jets. Concorde is supposed to be on the way in.

But noise was not the main thrust of the criticisms levelled by *Tomorrow's World* against Concorde. One of the country's largest research bodies half-jokingly produces its costings for big projects in terms of 'milli-Concorde' (milli-Concorde = a million pounds); half-jokingly, because none can quickly forget the drain Concorde represented on the nation's research and development spending over the past twenty years. The famous 'spin-off' argument is hard to sustain: *Tomorrow's World* looked long and hard at the suggested benefits which have come to Britain as a result of Concorde's development; apart from employment, it would be hard to value them at a ten thousandth of the final bill.

The figures are now very familiar. When the French and the British governments put their signatures to the famous concordat, the estimate stood at £150 million development costs to be set against sales of 160 aircraft. Two years later the first re-design had bumped up the estimated cost to £275 million and by 1966 that figure had reached £500 million: the figure had this time been boosted by 'development' difficulties. By now, some people were beginning to wonder whether the original estimates had been based on genuine costings or on a guess as to the figure likely to be accepted by the government of the day.

The planemakers assured the ministries that sales by 1980 were now estimated at 500, but by 1969 a second re-design of the wings and engine added a further £200 million to the cost, and now the Jumbo had arrived, eating into Concorde's likely routes; sales estimates were halved to 250. From then on the rot set in. Development costs were jumping at the rate of £75 million a year; by 1974 Concorde was six years behind schedule, only sixteen were being built, and the cost had reached the £1000 million mark.

That development cost has now been written off (it does not sound quite so large when described by the MP as £30 million a year spent by Britain over the fourteen-year programme) but there was more to come. To build the sixteen Concordes was going to cost a further £500 million, and this money, in theory, was to be recovered by selling the planes. In fact, British Airways had bought five by 1978 at a cost of £160 million, including six replacement engines; Air France too had bought five; two were by then museum pieces and two more 'on offer' to any other airline which would be prepared to buy them. The question facing the airlines is whether they can recover both the purchase price and the running-costs of their supersonic airliners.

One stark fact about Concorde is that a very high percentage of the weight she carries on take-off is fuel: a typical breakdown is 90 per cent fuel, 10 per cent payload (the Jumbo's ratio is 70/30). That, estimated British Airways, should carry 100 passengers (not 120) for 3500 miles under ideal conditions. Given a longer route, or less than ideal conditions, the equation which now faces the airlines is that for every 1 per cent extra fuel carried, ten passenger seats have to be left empty.

Now, when Concorde flies below the speed of sound – which she has to do over almost all populated countries – the aircraft actually uses more fuel to cover the same distance than when flying supersonically. That is why the New York route is so favourable to Concorde: a 3500-mile route mainly across the sea and much used by businessmen. Since Kennedy Airport has been opened to Concorde, the payload has averaged seventy-nine passengers, nineteen more than needed to cover the actual operating costs (but not to repay the purchase price). The route has proved so successful that in the summer of 1978 the number of flights in each direction has been boosted from seven to ten a week.

The other route on which Concorde was due to operate was less ideal. Bahrain had admittedly been chosen only as a staging post on the important route to Singapore and possibly onward to Australia, but it proved to be a journey which showed up the shortcomings of Concorde's operating economics. The first stage of the journey is overland across Europe to Trieste, and this stage had to be flown subsonically. Bahrain itself is hot, often searingly hot, and this imposed two limitations. The heat lessened the strength of the tyres, reducing the number of passengers which could be carried, and take-off was also affected by the fact that the blazing sun prevented the engines developing full thrust. At the time British Airways reckoned that the payload to Bahrain would be ninety-five passengers, which was disappointing for a leg of only 3100 miles, but on the return journey the heat reduced the number of passengers to a mere seventy.

Two years later, the record of the Bahrain route has been even more disappointing than that. The weekly journey has been made by, on average, only thirty people; that is how few travellers want to go to Bahrain, and both the Indian and the Malaysian governments have been obdurate in preventing Concorde from flying over their countries on her way to Singapore. (One notable exception to this general reluctance has been a Leigh-on-Sea architect who.has so far flown seventy times to and from Bahrain at a cost either to himself or his Arab employers of £25,000!) But there has been some good news on the technological front; one rubber firm has improved the strength of the rubber in the tyre wheels enabling a take-off speed which is five knots higher than in 1976. This, plus the development of a new High Specific Gravity Fuel, means that Concorde can now take off from Bahrain in high temperatures with as many as eighty-nine passengers; if only that many wanted to make the journey.

British Airways doggedly insist they can make money on Concorde eventually. Assisted, of course, by the fact that they are the sole operators of the SST on the routes they fly, they point out that if they can boost the number of hours flown by each aircraft to 2750 a year, carrying sixty passengers on average on every flight, they will start to claw back the overheads, particularly towards the end of what they hope will be a full ten years' flying life for each Concorde. They have signed an agreement with Braniff to lease the aircraft to the Texas airline every time Concorde reaches Washington, so that Braniff can operate a service onwards to Dallas/Fort Worth and back. They have at last won approval for that Singapore routing, and console themselves that they are the possessors of the most expensive status symbol in the world.

There is no pleasure in reporting the failure of this massive venture, the loss of £1000 millions to Britain and France, and the loss of 2000 jobs in England and more in France. Can anything be salvaged from the wreckage? Bob Hage, marketing chief of America's McDonnell Douglas Corporation, believes that the 1990s will see a new generation of 250-seater SSTs, and he claims that Britain's know-how will be crucial to this venture. Will it be possible to persuade the British taxpayers, and the scientists whose research budgets have suffered once on behalf of mega-projects like Concorde, that they should this time believe the optimistic projections and cost estimates of the aerospace enthusiasts? If *Tomorrow's World* was born in the sixties – the years of the 'white heat of technology' – the Concorde programme matched exactly the mood which has since come over Britain, the feeling that the sixties had been a decade of scientific and technological extravagance, the seventies of stringency and reappraisal, and as for the eighties. . . ?

Whatever happened to . . . Moulton's Coach?

After four tales of plucky failure I had hoped to end with a success story, even where it proved a *Tomorrow's World* prediction to have been wrong.

In 1975, the year in which a great deal of publicity had been attracted by a number of coach accidents in which all too many passengers had been killed, unprotected by the flimsy structure of most British coaches, the programme demonstrated a prototype vehicle which not only offered a high degree of roll-over and side protection, but at the same time gave passengers the comfort of a luxury car and a very stable ride. We ended the report with the following words:

'So when are we going to see these coaches coming off the production line? Well, the sorry answer is, possibly never, in this country anyway. The leading British coach manufacturer, after showing a great deal of interest, has finally withdrawn, and no other British coach manufacturer has yet taken up the design. So it looks as if we could be seeing yet another chapter in what seems to be the British disease: an excellent prototype which never gets into production. Meanwhile, here in Bradford-on-Avon the design team sits, and waits, and hopes.'

Head of that design team is one of the few designers in Britain who can claim to be a household name, Alex Moulton, designer of the hydrolastic suspension for Issigonis' famous and still successful Mini, responsible also, and more famously, for reducing the size of bicycle wheels, the first radical change in cycle design since the Victorians climbed down from their penny farthings. But Moulton has by no means retired to live on his well-deserved earnings from these two ventures. In his almost stately home in Bradford-on-Avon in Wiltshire, Moulton heads a team of designers and technicians seeking, among other things, to find fresh uses for their fast-growing expertise with rubber.

It was Moulton's success with the hydrolastic suspension of the Mini which started the ball rolling with his biggest baby, the coach. The beginning came when Sir John Harriman, Chairman in the late sixties of the British Motor Corporation, suggested that Moulton follow his success with cars by adapting the hydrolastic suspension to trucks. Moulton was sceptical that truck drivers would be the best market for such a luxury ride as hydrolastic could afford, but nonetheless, with funds from BMC and Dunlop, the independent suspension was fitted to the chassis of a Thorneycroft Trident. The tests, ultimately unsuccessful, nonetheless taught Moulton one important lesson: on a large vehicle, an independent suspension needs to be fitted to an exceptionally rigid frame. The free ride of the wheels subjects the body of the vehicle to more stresses and strains than conventional suspension systems.

If truck drivers are not a prime target for a comfortable ride, then coach passengers surely are, and Moulton switched his attention towards this market. He was still beholden to BMC, but the time was fast approaching when Leyland was to take the corporation over; Dunlop had always backed the venture, and it is behind it to this day. Aware of the need for that rigid frame, Moulton came up with a geodetic structure, a veritable honeycomb of welded steel tube, relatively light but immensely strong. It had the additional bonus that it provided protection for passengers should the coach roll over – most British coaches consist merely of a flimsy superstructure of soft aluminium and glass built on top of a standard chassis. Moulton's design integrates the frame of the body with the chassis itself.

He decided on no fewer than eight wheels on four axles, which effectively meant that, however bumpy the road, there were always likely to be six wheels riding level. The wheels he chose were small by coach standards, a design feature which later came in for criticism but which gave the coach a very striking appearance. By 1969, Moulton had completed his design and invited British Leyland down to inspect it.

They came to Bradford-on-Avon in strength: Lord Stokes, Lyon, Fogg and Ellis came and tested the prototype with what appeared to be intense interest but, alas, they were not impressed. The Leyland Truck and Bus division designers say today that the main factors which put them off were the small wheels (impossible to supply 'off the shelf', too low for ground clearance in rough country, and the tyres wear out rapidly), the lack of luggage space, and the fact

As comfortable at speed as a luxury limousine, with its eight wheels riding easily over any potholes or bumps in the road, the Moulton Coach cruising at speed

that Moulton had designed his coach around a very varied set of components, not relying on any one supplier. Their reaction was also no doubt influenced by the fact that Leyland had just completed negotiations for a large government funding for its own Leyland National Bus, again a very well-built, safety-conscious coach, and one which relied very heavily on new investment in production machinery. Moulton's low-capital construction would hardly have fitted in with Leyland's game plan! So, no villains in the piece, but no immediate hope for Moulton in getting his coach on the road. Leyland did undertake to help in the search for a smaller company to manufacture the coach, but with no success.

In the meantime, the trade press continued to shower praise on the comfort and the handling offered by the prototype; *Commercial Motor* compared it favourably with the ride of a Rolls-Royce Silver Shadow, *Motor Transport* contented itself by saying it was as good as the Advanced Passenger Train. When *Tomorrow's World* rode over a very bumpy test track the eight independently suspended wheels rode serenely over any kind of pothole; William Woollard found that driving it was a pleasure, and the braking power remarkable.

All of which impressed other would-be customers. Volvo came and professed itself 'wildly enthusiastic'. Alex Moulton in fact believed that the Swedes had agreed to buy the design, but when they got back home the mood changed abruptly; the motor industry in Sweden was just about to enter a period of crisis, and the alarm bells sounded just too soon for the deal to be clinched. The gloomy comment which ended the *Tomorrow's World* report reflected pretty accurately the mood of 1975.

But not long after the coach appeared on television, Moulton was contacted by the Hestair Dennis group – an amalgam of the specialist motor makers, Dennis, with the larger Hestair company. It was for the Dennis workshops in Guildford that a deal was soon struck; a few design modifications were made: the wheels were now larger, there was more luggage space, and the overall shape was a little less angular than the original, but the main thing was that there was now a real chance that Moulton's coach would be seen on the roads. Moulton's insistence that the coach would be – apart from the suspension stations – easy to assemble from off-the-shelf components had paid off at last. Even the coach builders have relatively little to do, because the intricate tubular body frame takes most of the expense out of this part of assembly. Moulton in fact believes that the coach could be sold overseas in do-it-yourself assembly kits, comprising basically the four suspension units and the frame, and that any developing country could build a coach industry around them, using only the simplest of rigs and relatively unskilled labour. (One of the early tests was an assembly using thoroughly bad welding; the frame turned out to be perfectly functional.)

In the summer of 1978, the first full-scale model of the new design was put into wind tunnel tests at Bath University, and by early 1979 the first two coaches would have been ready for endurance testing. But despite the fact that the coach had emerged triumphantly from the wind tunnel tests, Hestair Dennis considered that the coach could not be manufactured profitably and decided not to put the coach into production. Moulton is talking hard with yet another potential buyer.

'How many of the items you show on *Tomorrow's World* turn out to be successes?' Not many, especially if they are prototypes.

The frame of the Moulton Coach not only provides passengers with a unique degree of safety should the coach roll over, but it is designed for easy assembly from kit form by any low-technology overseas company

Picture Credits